The Science of Space-Time

Derek J. Raine

Leicester University

and

Michael Heller

Pontifical Faculty, Cracow and Vatican Observatory, Castel Gandolfo

Pachart Publishing House

Tucson

Library of Congress Catalog Card Number: 81-83129
International Standard Book Number: 0-912918-12-8

In this book the typescript prepared by the author has been directly reproduced in its original form. This method permits this volume to be available as economically and as rapidly as possible. It is hoped that necessary typographical limitations of this method will in no way distract the reader.

Pachart Publishing House
1130 San Lucas Circle
P.O. Box 35549
Tucson, Arizona 85740

Table of Contents

Preface

- Nevertheless, it moves.
- But it is a matter of great difficulty to discover the *true* from the apparent motions.
- The universe is not *twice* given, with an earth at rest and an earth in motion; but only *once*, with its *relative* motions, alone determinable.
- Henceforth there should be no inertia of bodies relative to space, but only inertia of matter relative to matter.

The history of the problem of motion is at the same time the history of the nature of space and time. Space and time enter into physics in a way that is beyond our control. We are not free to repeat an experiment at exactly the same time and at exactly the same location with respect to the stars. We are not free to adjust the structure of space-time to suit our *a priori* intuitions or prejudices. For the structure of space-time is revealed to us in the construction of our theories of motion and in their relation to observation. The appreciation of the significance of this result is relatively new. It was discovered for us by Einstein, not once but twice: in the special and general theories of relativity. In the special theory the structure of space-time in the absence of gravitational forces is revealed to us through the motion of light. In the general theory the motion of bodies under gravity provides us with a more complete picture. And in the general theory the space-time structure itself is governed by laws of motion, and thereby becomes the subject of a dynamical theory.

The purpose of this book is the analysis of the development of the structure of space-time from the theory of Aristotle to the present day. The concept of relativity may be thought of as entering in the way in which space-time splits up into space and time. And it enters also through the ideas that Einstein called 'Mach's Principle', concerning the relativity of inertia.

Although we raise and discuss many philosophical issues, this book is not intended as a text in philosophy, but rather as a prelude and preparation for philosophy. It is not a history of the subject, since we have made no attempt to be complete, emphasising, rather, with considerable hindsight, the major steps in the development. Nor is it intended as a complete introduction to the general theory of relativity, since our purpose is to analyse how things fit together, not to teach how calculations can be carried out. There are many excellent introductory texts available and in continuous creation.

Our hopes for the role for this book might have been best conveyed by calling it 'The Science of Relativistic Mechanics', but to have invited such direct comparison with Mach's masterpiece would not have adequately expressed our fears for the limitations of our worth.

Much of the book should be accessible to readers with only a modest mathematical background. It is impossible to avoid mathematics entirely without an incomprehensible abundance of circumlocution. And it is difficult to write mathematics which avoids the for us often unnecessary precision of mathematical notation. Many of the formulae that we require for precision can be read simply as expressions of the existence of determinate relationships between the stated quantities. There is, however, an unavoidable progression in mathematical level through the book. Some details are commented on in the Appendix, although for the most part we do not dwell on technicalities, all of which can be found in many standard references.

We have, of course, benefitted from discussions with many colleagues. In particular, Dr. D.W. Sciama has been above all a source of guidance and inspiration. Comments of Dr. J. Barbour influenced our discussion, particularly in Chapter 4 and in Section 11.3. D.J.R. would like to thank some of his colleagues at the Open University for raising awkward questions while unaware of precisely why they were so encouraged to provide also the answers.

But while others may have been responsible for much that went
into our heads, they are in no way to blame for how it has come
out.

One of us (M.H.) would like to offer his appreciation for
hospitality at various stages during the writing of this book to
Professor O. Godart of the University of Louvain-le-Neve, to Professor Blackwell of the University of Oxford, and to Professor
Meadows of the University of Leicester. One of us (D.J.R.) would
like to thank his wife for hospitality during the long period
when his thoughts were clearly elsewhere. And we are grateful to
our typist for dealing so uncomplainingly with a manuscript that
must have appeared to have undergone its final revision only when
its authors found that they themselves could no longer decipher
it, and which unfailingly failed to materialise on time.

Chapter 1: Aristotelian Dynamics

The history of our understanding of both geometry and dynamics can be traced back at least to classical Greece, but the
development of the conceptual framework under which these two
branches of thought could be brought together occurred only at
the beginning of our own century. Nevertheless, we can use this
conceptual framework to illustrate, with considerable hindsight,
the history of our understanding of the structure of space-time
as a process of continuous evolution. If the principle of this
evolutionary process can be summarised in a single thought it is
that the structure of space and time, far from being a free
creation of the human intellect, is constrained for us by the
observed behaviour of the material world. It is conceivable that
the ultimate goal of this development involves the complete
determination of the structure of space-time through matter, a
synthesis to which our present theories do not yet aspire but may
point the need.

Our approach is not strictly historical since one cannot
measure the struggles of the past with the standards of achievement of the present. Those generations on whose shoulders we
stand sometimes lacked the conceptual language with which to ask
the right questions of a dynamical theory, or of the structure of
space; we shall in part supply this. Conversely, the conceptual
puzzles of yesterday do not always exist in the language of today,

1

and we shall not always be concerned to resurrect them. The
appropriate language is drawn to a large extent from recently
evolved concepts of mathematics, although not necessarily from
its symbolism, and is to that extent technical. In some cases
our use of contemporary language may appear to complicate the
picture, but it is intended to enable us to elucidate clearly the
relevant issues. We aim to rationalise the past from the standpoint of the present and, by using a common language, we shall
see how the modern view of space-time structure builds on, contrasts with, and has evolved from its origins in antiquity.

Our discussion of the evolution of dynamical thought
begins with a paradox which, in the form of Zeno's arrow, appears
to deny the possibility of motion. This may seem an unlikely
start to dynamics, but it is no stranger than the origin of
Einstein's special theory of relativity in his paradoxical
attempts to visualise the world of an observer riding on a beam
of light; or of the beginnings of the quantum theory of matter
in Rutherford's experiments on the structure of atoms which
showed that those atoms could not exist. These examples have in
common the subversion of an inadequate conceptual framework
through an essentially simple observation of the world. The
paradoxes define the boundaries beyond which received intuition
cannot be pushed, and the possibility of progress with confidence
has to await their solution. Thus it is only with hindsight that
we are able to place at the beginning of dynamical theory those
laws which Aristotle proposed to describe the motion of bodies.
For these laws could be of little practical consequence until
Newton and Leibniz had invented, in the calculus, a machinery
for circumventing Zeno's paradoxes.

§1.1   *Local and global concepts of motion*

The motion of a body can be looked at from two points of
view: either locally (differential viewpoint), point by point
and from moment to moment;   or globally (integral viewpoint), as
a path from an initial to a final point. Without the tools of
infinitesimal calculus the local approach is doomed to inaccuracy
and paradox. This is illustrated in the antinomies of Zeno of
Elea.

According to Aristotle in the antinomy of Achilles and the
tortoise, Zeno

> purports to show that the slowest will never be
> overtaken on its course by the swiftest, inasmuch
> as, reckoning from any given instant, the pursuer,

2

before he can catch the pursued, must reach the
point from which the pursued started at that
instant, and so the slower will always be some
distance in advance of the swifter. (VI Phys. 239b)

The paradox arises from the infinite divisibility of space and
time. At each successive step, Zeno invites us to contemplate
the diminishing distance covered, but neglects the fact that
these distances are covered in diminishing times. The race comes
to an end when both become zero; Zeno's argument shows only that
it does not come to an end before this. By summing a geometric
series of infinitely many terms we find the duration to be finite.
There is no contradiction between the infinity of terms in the
series, which thus appears to go on forever, and the finite value
of the time taken to generate the series, although the properties
of space and time which we need to ensure this absence of
contradiction are, as we shall see, by no means trivial. From a
global point of view a finite path is traversed in a finite time
at a constant velocity. By infinite division of the path we make
contact with the local viewpoint: successive divisions of the
relative distance and of the time of travel tend individually to
zero, but their ratio tends to a finite number, the velocity of
Achilles.

In a sense here the tortoise is redundant and serves only
to complicate the problem. From the point of view of the
tortoise, that is, from a frame of reference in which the
tortoise is at rest, Achilles' task is to cover a certain
distance at a certain (relative) velocity. If Zeno were right
and Achilles were not able to perform this task then the argument
would show that no motion at all is possible. Indeed in the
antinomy of the arrow Zeno attempts to show the impossibility of
motion directly. Here he claims that

since a thing at rest when it has not shifted
in any degree out of a place equal to its own
dimensions, and since at any given instant during
the whole of its supposed motion the supposed
moving thing is in the place it occupies at that
instant, the arrow is not moving at any time during
its flight. (VI Phys. 239b)

The problem is again in the infinite divisibility of space and
time as we concentrate on smaller and smaller displacements in
shorter and shorter times. At any instant it is perfectly true
that the arrow can be considered to occupy a definite position.

3

This is assumed in setting up classical dynamics, where we
identify a position $x(t)$ as a function of time t, meaning thereby that a unique x is to correspond to a given t. We can define
the velocity at any instant as the ratio of successively smaller
increments in position to the corresponding times of flight, and
it does not follow that the ratio tends to zero along with these
increments. Consequently there is no contradiction in classical
dynamics in the assignment of a definite position and velocity at
any instant.

Neither the concept of a mathematical function approaching
a limit nor the idea of continuity were available until modern
times. The relation between local and global concepts of motion
could not therefore be elucidated. It is no trivial matter to
construct a mathematical model for space and time which is
satisfactory for the description of the motion of bodies, that is,
for kinematics.

§1.2   *Space-time models for kinematics*

Between any two instants of time we can find a further
instant; between any two positions in space there is at least
one location. If therefore we want a kinematical model in which
a body can have a definite position at any instant, we must incorporate this property of infinite divisibility. From the set
of integers by means of simple proportion we can construct the
set of rational numbers. That these do not yet have the desired
property is, of course, the famous problem of irrationals in
Greek mathematics. This problem was solved by construction:
from the rationals the 'continuum' of real numbers R is constructed by means of the Dedekind cut (Dedekind, 1901). In this
way the properties of the real line are not given but arrived at
by construction. The Dedekind cut constructs only certain
properties of R, namely the continuum or infinite divisibility
property, in addition to the algebraic properties of real numbers.
It does not yet, for example, endow R with a 'metric', that is,
with a means of computing the distance between points of R. This
is a property which must be constructed separately as the metric
geometry of R.

From the one dimensional space, R, of points, (x), we can
construct the three dimensional continuum $R^3 = R \times R \times R$ of
ordered triples (x,y,z). This will serve as our model of space
for the present. As our model of time we take the real line R
itself. Since this model will turn out to be inadquate in many
ways, these remarks are not mere pedantry.

We can now construct paths of particles in space as

mappings from R to R^3: $t \rightarrow (x(t), y(t), z(t))$ which we shall
write more compactly as $t \rightarrow \underset{\sim}{x}(t)$.

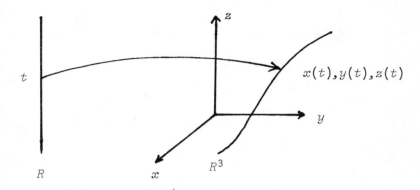

Fig. 1.1.   The map from R to R^3 is a particle path in R^3.

Although we have not yet specified a distance function on R, so
we cannot say *how* close to each other the points of R are, we can
use the usual order relation on R to determine which of any two
points is closer to any third point.   Consequently, we know from
the order relation if a point P_3 is between points P_1 and P_2
since, if so, P_3 will be closer than P_2 to P_1.   We can call a set
of points 'open' if (a) between any two points of the set there
is at least one further point of the set;   and (b) any point of
the set is between at least two other points of the set.

Technically, by specifying which sets are to be called
'open', we are giving R a 'topology' (or giving it the structure
of a topological space).   Thus the usual order relation amongst
the real numbers gives rise to one possible topology for R.   The
crucial point is that the topological structure is required to
enable us to define 'continuous functions' f: $R \rightarrow R$.

Intuitively, a function $y = f(x)$ is continuous if its
graph (x,y) in $R \times R$ can be drawn without removing pen from paper.
Technically, this becomes the requirement that, at least
sufficiently close to a point at which the continuity of f is
under investigation, the inverse mapping f^{-1}: $y = f(x) \rightarrow x$
preserves the open sets of R.   By this we mean that if X is an
open set of points in the range of f then so is $f^{-1}(X)$ (Fig. 1.2).

Finally, we are able to define *continuous* particle paths
in R^3 by specifying that a path $\underset{\sim}{x}(t)$ is continuous if each of the
three coordinate functions $t \rightarrow x(t)$, $t \rightarrow y(t)$, $t \rightarrow z(t)$ is con-

5

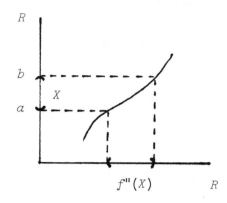

Fig. 1.2.   A function f is continuous if f^{-1} maps the
open set $X = \{x\,|\,a{<}x{<}b\}$ in the range of f
into the open set $f^{-1}(X)$.

tinuous as a mapping from R to R in the above sense. This construction corresponds to our intuitive ideas of continuity since
these are derived from three-dimensional Euclidean geometry. It
is possible to have different topologies for R^3 in which
different paths would be regarded as continuous. It is not impossible — and we would not wish to make this point any more
strongly — that an alternative topology might provide a better
model for the space and time of the microscopic world than that
of our 'common sense'.
        The structure we have so far described is still not suff-
icient for Aristotelian dynamics since it does not yet enable us
to construct a velocity vector. This is because a path may be
continuous without possessing a tangent vector representing the
velocity at any point (Fig. 1.3). Since we require our particles
to have velocities we need to distinguish those paths which have
this additional property. This is possible since R^3 has the
structure of a 'differentiable manifold'. This means here simply
that we know which functions on R^3 can be differentiated, since
we simply refer to the usual rules of differential calculus to
adjudicate any particular case.
        To see that a knowledge of the differentiable functions is
required to define a tangent vector is a little difficult,
depending as it does on some precise definitions. It is necessary
to distinguish between the set of tangent vectors at a point,
which depend on the differentiable manifold structure and are not

6

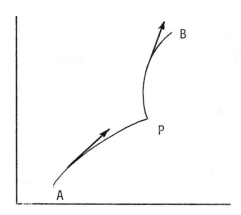

Fig. 1.3.   The continuous path APB has no tangent vector at P.

a priori   defined to belong to the space itself, and the notion
of R^3 itself as a space of vectors, or directed line-segments,
emanating from the origin, a concept which does not depend on
differentiable structure.   The difficulty is compounded by the
fact that in R^3 (or, in general R^n) the two structures are
'isomorphic', which means that *once constructed* they can be taken
to be the same.   It is only in the construction of the structures
that the difference arises.   The connection between differenti-
ability and tangent vectors can be made plausible if we recall
that the derivative of a function is a measure of its slope at a
point, and that slope can be represented by a vector in the sense
of a directed line-segment (of arbitrary length).   Collecting together all the different slopes of different functions therefore
generates all possible vectors at a point.   Note that we do not
yet specify a way of measuring the length of vectors — this is
yet another separate structure, a metric, which we impose
separately.

Our discussion is clearly much more general than the simple
idea of a vector as a directed line-segment.   The point of importance is that we see how the introduction of vectors — hence
the velocities of particles along their paths — depends on the
structure of space-time as a differentiable manifold, while the
trajectories themselves require only the idea of continuity.   By
means of the manifold structure of space, we can give a complete
kinematical description of motion and establish contact between
the global and the differential point of view.

This is not yet all the structure with which we intuitively

7

endow space. An observer is able to compare the velocities of particles at different points: it makes sense to speak of the velocity of the runners in a race as being different even before we add the structure necessary to quantify that difference. The possibility of comparing vectors at different points arises because we are able to transfer vectors from point to point by parallel displacement. The existence of an operation of parallel displacement is called an 'affine structure'. This structure is again to be found through experiment. It will not necessarily turn out to be the 'obvious' one that can be constructed from the intuitive concept of parallelism.

Finally, we have a way of quantifying intervals of space through measurement by rulers, and intervals of time using clocks. Good clocks are supposed to run smoothly in the sense that at any epoch the measure of the interval between times t_1 and t_2 is $(t_2 - t_1)$. This is not mandatory. For example, in the musical scale the measure of an interval is not the difference in frequencies but their ratio, hence the difference in the logarithms of frequencies. Consequently, the measure of the frequency interval depends on register. Correspondingly, the measure of time interval could be some non-linear function of t_1 and t_2, and, indeed, will be in the case of badly constructed clocks. In order to employ this 'Euclidean metric' of the real line to respresent time intervals, we assume the existence of some standard clock (e.g. atomic clocks).

The distances in space we again assume to be given by the Euclidean metric; Pythagoras' theorem then gives us the distance $d(\underset{\sim}{x}, \underset{\sim}{X})$ in R^3 between points (x,y,z) and (X,Y,Z) as:

$$(d(\underset{\sim}{x}, \underset{\sim}{X}))^2 \;=\; (x - X)^2 + (y - Y)^2 + (z - Z)^2 \;\;.$$

This distance is supposed to be measurable by means of some standard ruler, although not necessarily directly or in any simple way.

The existence of a metric structure enables us to quantify velocities. This appears intuitively obvious since we think globally of motion over a certain distance in a certain time. However, there is a slight subtlety here. The global view is strictly appropriate only to the manifold R^3 with its Euclidean metric. As we have indicated, only for manifolds of this type can we think of a velocity vector as an arrow between two points in space (viz. those separated by the distance travelled in unit time) to which we can apply Pythagoras' theorem. In other cases we shall have to be a little more careful in our discussion of the metric (see Sect. 7.3).

8

To pass from a kinematical description of motion to a dynamical one, we introduce the Aristotelean analogue of Newton's first law, the 'law of inertia':

> *First Law:*   if an object is in motion it is
> necessarily being kept in motion
> by a cause.

This law implies the existence of a state of absolute rest, which is the state of a body subject to no cause of motion. We shall see that this implies that the Law has a fundamental role to play in determining further the structure of space and time, a role which is in strict analogy with that played by Newton's first law.

Aristotle's use of the term 'cause' was wider than the modern meaning. We use the term for what Aristotle would call the 'efficient cause'. Enforced motion, which by definition means departure from 'natural' motion, required an efficient cause for its initiation and continuation. Natural motions required a 'final cause', or purpose, which was provided by an 'unmoved mover'. The first law includes both types of cause.

As to the relation between a force and the enforced motion it produces, Aristotle writes:

> If, then, A is the moving agent, B the mobile,
> C the distance traversed and D the time taken,
> then A will move ½B over the distance 2C in time
> D, and A will move ½B over the distance C in time
> ½D;   for so the proportion will be observed.
>   (VII Phys. 249b-250a)

It is reasonable to interpret this to mean that the 'force' exerted by A is proportional to the mass of B, to the distance covered C, and inversely proportional to the duration D. In modern notation, writing F for the force and m for the mass of B, we have the

> *Second Law:* $\underset{\sim}{F} = m\underset{\sim}{v}$, where $\underset{\sim}{v} = d\underset{\sim}{x}/dt$ is the
> instantaneous velocity.

This is in agreement with the first law since it gives $\underset{\sim}{v} = 0$ if $\underset{\sim}{F} = 0$. As we shall discuss more fully in connection with

Newton's Laws, the second law does not make the first redundant. Essentially, this is because the first law states the existence of a certain class of motions and space-time structure in terms of which the second law acquires its meaning. It is amusing to note that Aristotle did not entirely believe the consequences of the second law: he argues that if a force is very small in relation to the mass it cannot produce motion because 'if it were so, then a single man could haul the ship' (VII Phys. 250a). (The resolution of the paradox is, of course, the existence of static friction.)

§1.4   *Aristotle's cosmology*

In order to give an empirical meaning to the two dynamical laws it is necessary to be able to identify a state of rest, not merely to assert its existence. In Aristotelean dynamics it would be circular to identify such a state as the absence of forces, since in the case of natural motions the required forces are defined through the observed motions. Thus any motion would be possible by asserting it to be natural, and the dynamical equations would be empty. What is required, therefore, is either an independent identification of natural motions, so that an observer of these motions as prescribed acquires a privileged status with regard to the dynamical laws; or, alternatively, the direct identification of a state of rest. Through his cosmology Aristotle provides both.

It has become customary to refer to a statement on the position of the Earth in the Universe as a 'Cosmological Principle'. One might therefore call the 'Ancient Cosmological Principle' the assertion that the Earth is the immovable centre of the Universe. Having dismissed the heliocentric system, this is the position Aristotle adopts. Consequently, all motion is to be referred to the Earth as an absolute standard of rest, and all location is to be referred to the Earth as an absolute centre. It is motion relative to the Earth that requires a force.

Aristotle now goes on to distinguish the natural motions. For this, he divides the heavens into a sublunar sphere and a region beyond the orbit of the Moon. The latter consists of fifty-five spheres, centred on the centre of the Earth, on which the heavenly bodies move in perfect circular motion, and is closed by the sphere of 'fixed stars'. In contrast, the natural motions of the sublunar world are rectilinear, towards, or away from the centre of the Earth.

This set of motions uniquely defines a privileged observer. Such an observer is at rest at the centre of the Earth;

10

for no other observer are the orbits of heavenly bodies as described. For such an observer the laws of dynamics are supposed
to be valid, the velocities therefore being velocities relative
to the Earth. In this way Aristotle's two prescriptions are
consistent, and his dynamics can be subjected to empirical test.
Of course, the laws are now useless for predicting anything
concerning the motion of the heavenly bodies, since the forces
acting on them are defined by their motion. On the other hand,
the forces which produce departures from natural motion are to
be thought of as having a meaning outside the motion they generate. These forces, which are for the most part electrical in
origin, can be measured, at least in principle, in static
situations. Had Aristotle attempted to do this he would have
found that his laws did not fit the observations. It is important to emphasise that Aristotelian dynamics fails as a
description of Nature not because of any inherent logical flaw,
but simply because it does not produce results in agreement with
observation.

§1.5 *The Mach problem in Aristotelian dynamics*

There is however a problem which has more of a philosophical than empirical character which is common to Aristotle,
Newton and Einstein, and which, following Einstein, is discussed
under the name of 'Mach's Principle'. As far as Aristotelian
dynamics is concerned the problem is as follows. The dynamical
laws involve absolute not relative velocities. It is not sufficient to know that two bodies are in relative motion; one
must know which, if either, is absolutely at rest, or how
rapidly both bodies are 'really' moving. Consequently, there is
a need to introduce an absolute space, a backcloth against which
the motion of bodies is to be described according to the laws of
dynamics and the accidents of initial conditions. If we view
the motion of bodies from a frame of reference moving with an
absolute velocity $\underset{\sim}{u}$, then the second law governing the motion of
a body observed to have relative velocity $\underset{\sim}{v}$ is

$$\underset{\sim}{F} = m(\underset{\sim}{u} + \underset{\sim}{v}) \; ,$$

since the absolute velocity is $\underset{\sim}{u} + \underset{\sim}{v}$. We can write this as

$$\underset{\sim}{F} - m\underset{\sim}{u} = m\underset{\sim}{v} \; .$$

In this form we can see an additional 'force', $- m\underset{\sim}{u}$, has to be
introduced into the equation of motion. In the moving frame

Aristotle's law is not valid unless we include this 'force' which is supposed to arise from motion relative to absolute space. Now it is possible to identify absolute space as that frame of reference at rest at the centre of the Earth, as Aristotle indeed does. In this way the purely operational problem of applying the law is removed. However, one cannot dispose of the whole problem in this way. For there is nothing in the nature of the inertial force to relate its existence to motion of frames of reference relative to the Earth; there is, for example, no causal connection. The existence of absolute space and the absolute stationarity of the Earth are logically distinct elements of the theory. Since neither is a consequence of the other, the agreement of frames of reference attached to the Earth and to absolute space, far from solving anything, becomes a remarkable coincidence in need of explanation.

§1.6   *Aristotelian space-time*

In order to bring out the full structure of Aristotelian dynamics, and to see its relation to subsequent developments, it is useful to put the theory into a 'space-time' setting. The idea that 'space-time' rather than space and time separately is the arena for dynamics was first introduced by Minkowski in connection with the Special Theory of Relativity. However, the concept is applicable also to prerelativistic systems, various theories differing in adopting different structures for space-time.

Instead of representing paths of particles as curves in R^3 parameterised by a real parameter t, we consider the path in R^4 in which the position and time coordinates are functions of a parameter λ. Such a path is called a 'world-line'. A point on a given world line, corresponding to a particular value of λ, represents the position $x(\lambda)$, of the particle at the time $t(\lambda)$, that is, an 'event' in the particle's history. No physical significance need be attributed to the parameter λ.

We now have to see what the dynamical laws tell us about the structure of the space-time manifold M. We already know $M = R^4$ from our previous discussion, but this does not yet completely characterise the structure of M. Let us see how this can be determined for Aristotelian space-time. First, we notice that the second law requires the existence of an 'absolute' time: the *form* of the law will be altered if we choose a new time $\tau = f(t)$ since

$$v = \frac{dx}{dt} = \frac{dx}{d\tau}\frac{df}{dt} \neq \frac{dx}{d\tau}$$

12

unless $df/dt = 1$. The only permitted change is $f(t) = t + t_0$ where t_0 is a constant. This is merely a shift of origin. In fact, we are also permitted a change of units, $f(t) = kt$ with k a constant, provided we change the units of 'force' consistently. These transformations correspond to the obvious freedom to measure time from an arbitrary event, such as the birth of Christ or of Julius Caesar, and to measure it in numbers of years or seconds or coffee spoons (accurately calibrated). The existence of an accurate time-piece is postulated by the theory but its physical manifestation - if any - is a matter of experiment or observation, and, in the last analysis, convention. The permitted trans-formations are entirely trivial; only for an essentially unique choice of time could bodies be observed to move according to the laws of dynamics even in principle. Accordingly, to each event there is associated a definite time. More precisely, there is a 'projection map' π_T: $M \to T$ which takes any point x in the space-time manifold M into a point on a unique 'time axis' T, (which is just the real line). Since a time axis is singled out, coordinates in M are not on an equal footing: the space-time manifold has a decomposition into space and time determined physically by the laws of dynamics (Fig. 1.4). This expresses the existence of a concept of absolute simultaneity; two events can be meaningfully said to occur at the same time, no matter what their spatial separation may be.

Returning to the dynamical laws, we see that the first law requires the existence of absolute rest. This is supposed to be defined dynamically by the world-lines of bodies subject to no forces. Kinematically, it is identified by Aristotle as the ab-sence of motion relative to the Earth. Consequently, there is a projection map, π_Σ, of a point x in M into a three dimensional manifold, Σ, which is here R^3, our model for 'space'. The projection of events on the world-line of a particle at absolute rest is a single point of Σ, the absolute spatial location of the particle. The existence of this projection therefore allows us to talk meaningfully about the same place at different times. Notice that the existence of absolute time and absolute location (or space) are independent here, arising in fact from different dynamical laws.

The existence of these two projections is described by saying that the space-time manifold is a direct product of a one-dimensional (time) manifold T with a three-dimensional (space) manifold Σ, and we write for this $M = T \times \Sigma$. In identifying T with R we are assuming time to have no beginning and no end. As for space, Aristotle in fact assumes it to be bounded by the sphere of fixed stars, so that Σ is strictly a closed ball centred

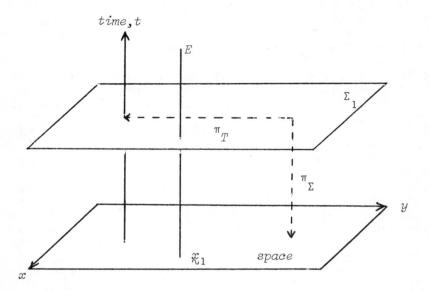

Fig. 1.4 The structure of Aristotelian space-time. Σ_1 is the
space of events at absolute time t, E is the
collection of events at absolute position x_1.

on the Earth. This is an unnecessary complication for us and we
can take Σ to be R^3.

In addition to the manifold structure, the geometry of T
and Σ is enriched with a Euclidean metric. The spatial separation between events at (t_1, x_1) and (t_2, x_2) is given by
Pythagoras' theorem,

$$|x_1 - x_2| = \{(x_1 - x_2)^2 + (y_1 - y_2)^2 + (z_1 - z_2)^2\},$$

and the time difference by $|t_2 - t_1|$, (which is the one-dimen-
sional form of Pythagoras' theorem). The structure of the space-
time is therefore a product of Euclidean spaces $E^1 \times E^3$. This
geometry has nothing to do with the motion of particles, but is
imposed independently, supposedly in accordance with the results
of separate experiments and measurements.

It is useful to consider the structure derived from the

14

existence of privileged trajectories of bodies at rest from an alternative point of view. At each point of space-time there will exist a vector of unit length tangent to these world-lines, which we can call the 4-velocity of the hypothetical body at that event. This accords with the terminology of special relativity and is to be contrasted with the 'ordinary' 3-velocity of the body which in this case is zero. We describe the existence of a unique vector at each point of space-time by saying that we have a 'vector field'. The existence of this vector-field is equivalent to the existence of absolute location.

This is not, however, the only privileged vector field in Aristotelian space-time. There are a further two which describe the natural motions. One of these is tangent to the trajectories of the points of the fifty-five spheres, and describes the uniform circular motion of the heavenly bodies. The other describes the natural motions towards or away from the centre of the Earth in the sub-lunar sphere. Strictly speaking we need several different vector-fields here to describe the motions of the different elements. The point that is important for us is that these motions are described by vector fields on the space-time manifold.

The use of so many unrelated vector-fields in Aristotle's dynamical theory may seem inelegant, as indeed it is when judged inappropriately in terms of the aesthetic criteria of modern science. Even so, it is not ridiculous. There are more modern theories which make use of more than one geometrical structure of a given type to describe related phenomena. Such are the theories of, for example, Milne (1935) and Dirac (1973). The difference lies in the fact that these later authors do not attempt to distinguish between phenomena on the basis of their proximity to the Deity, but on some more physical grounds; although it is possible that with hindsight the meaning of 'more physical grounds' in contemporary science may turn out to have a meaning similar to 'Deity' for the Greek philosophers.

Chapter 2: Copernican Kinematics

If, in the absence of a dynamical theory, the motions of the heavenly bodies can only be explained as 'natural' then they must be simple; for no complex motion could be satisfactorily considered as inscrutable and not requiring explanation. But as viewed from the Earth the planetary orbits are not simple and appeared to require a complicated machinery of circles moving on circles. The problem to which Copernicus was led therefore, was to reconcile the complexity of the observed motion of the planets with 'natural' circular motion in a way less artifically complex than these Ptolemaic epicycles. It is a matter of some debate whether he succeeded in this since the Copernican solution in detail is not markedly more efficient than the Ptolemaic. We know the reason for this is that we have to give up the idea of the circle as the only suitably natural orbit and that, as Kepler discovered, a simple description is achieved only if we are prepared to allow the planets to move on elliptical orbits. We know also, of course, that the solution of the problem of planetary motion requires a dynamical and not merely a kinematical theory. The problem of Copernicus and Kepler is not, however, without its modern counterparts. For example, we presume that the strong interaction, the force that holds the atomic nucleus together, should be treated by a dynamical theory, but we do not yet have that theory. A preliminary requirement is the systematisation of the interactions of elementary particles in terms of a 'natural' and 'simple' structure (which turns out to be the grouping into multiplets according to certain symmetries). If in our own day this is rated as a major advance, how much more so was the recognition of their problem, and its solution, by Copernicus and Kepler.

As in Aristotle, Copernicus allows himself just two types of natural motion, hence two privileged vector fields: straight lines for earthly bodies, circles for heavenly ones. This is quite reasonable if we picture the motion in space, rather than space-time, since the straight line is the archetypal motion without beginning or end, and the circle is the natural periodic path. Not until general relativity is it possible to see that both motions are 'natural' as part of a single space-time structure.

A word of caution should be added to this discussion. Fig. 2.1 illustrates the Copernican and Ptolemaic pictures of part of

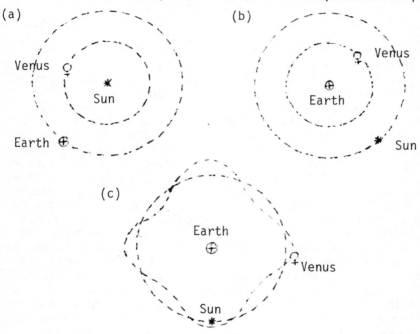

Fig. 2.1 (a) The Copernican Solar System. (b) The Ptolemaic Solar System. (c) A geocentric Solar System kinematically equivalent to (a).

17

the solar system as they are usually given (Fig. 2.1(a), 2.1(b)). As long as only the projections on the sky of the motion of the planets is under consideration both views can be made to fit the observations. Once distances from the Sun are taken into account the version of the Ptolemaic picture in Fig. 2.1(b) is demonstrably wrong and consequently allows no further discussion of much interest. However, one can still postulate a geocentric theory of the form shown in Fig. 2.1(c) in which distances are correctly represented, and which is, in fact, merely Fig. 2.1(a) drawn to accord with the viewpoint of an observer fixed on the Earth. In the subsequent discussion we are concerned with the comparison between the heliocentric view in Fig. 2.1(a) and the geocentric picture of Fig. 2.1(c).

§2.1   *The Copernican Cosmological Principle*

As a first step in the construction of a simple description of planetary orbits in terms of natural motions, Copernicus removes the Earth from the centre of the astronomical stage to a relatively lowly position amongst the other planets, and he sets the Sun in the central position. This is a first step towards the modern 'cosmological principle', which is indeed sometimes called the 'Copernican Principle'. This states very roughly that the Earth does not occupy a special place in the Universe. We shall have more to say about it later (Sect. 9.5). It is clear that the manifold structure of space-time has not been changed in passing from the Aristotelian geocentric universe to a heliocentric model. There is still an 'absolute place', since otherwise it makes no sense to discuss the location of the Sun without reference to an observer. The laws of natural motion are still to be given through a system of trajectories, or equivalently their tangent vector fields. These trajectories may be given either in space-time or in three-space, since from the spatial projections the space-time picture can be reconstructed. Since in both the geocentric and heliocentric theories the planets must follow the same paths, it would appear that the theories do not differ at all! The difference is, in fact, both important and subtle. According to the Copernican theory the laws of motion of the heavenly bodies amount to a prescription of certain fundamental vector fields. The fundamental fields are given in a special frame of reference attached to the Sun, and only in this frame do they take the simple form corresponding to circular motion. Indeed, this is the only frame in which the laws are given. In any other reference frame the vector fields are found by transformation and do not have a simple form. The object of

18

the exercise, indeed, is to find how certain linear combinations
of the vector fields appear to an observer on the Earth, these
linear combinations being tangent to the apparent orbits of the
planets. (The description by vector fields is introduced precisely because there is a natural way of combining vectors,
whereas the combination of trajectories is not so immediate.)
In the heliocentric theory the description of the vector fields
in any frame other than the special one attached to the Sun at
its centre involves velocities relative to the Sun. This singles
out the solar frame as special. In contrast, in the Aristotelian
(or Ptolemaic) description, it is the geocentric reference frame
that is privileged in this way.

This is an example of a point of some importance. We
start with a space-time manifold, with certain symmetry or 'invariance' properties. In this case it is R^4 in which the points
of the spatial sub-space R^3 are to be regarded as intrinsically
indistinguishable. At this level there are no special reference
frames. We then add physical or geometrical structures which we
suppose to be observable, in this case certain vector fields, in
relation to which all points of the space-time are not equivalent.
These structures define for us certain privileged frames.

Copernicus introduces a further set of vector fields to
describe natural motion in the vicinity of the planets. He argues
that the forces which hold the Earth in its spherical shape would
also so organise the Sun and the planets. Of course, this is
still far from Newton's universal gravity. Locally, the effect
is that the centres of planets become privileged locations, being
singular points of the vector fields. Here again, the appearance
of privilege is related to a physical observable, the distribution
of matter.

A theory will be said to be 'covariant' if the objects and
relations of the theory satisfy well defined transformation laws
which enable them to be described uniquely and consistently in
arbitrary frames of reference. The existence of special reference
frames is not sufficient to destroy covariance. For example, the
fundamental vector fields here are good geometrical objects which
can be described by any observer in any reference frame. Nevertheless, in a general frame these fundamental fields, hence the
natural laws, involve an absolute velocity. The meaningfulness
of the theory is guaranteed by the physical identification of
this velocity as a velocity relative to the Sun.

There is a contrast here with the use of the Copernican
principle in modern cosmology (Sect. 9.5). Here too the absence
of privilege for the Earth leads to the introduction of special
reference frames defined observationally in terms of the matter

in the Universe. However, the laws of nature are postulated
before the introduction of these special frames and are supposed
to apply whether or not these special situations exist. In this
way, one is led to believe that what is being described in our
physical laws is not one World but a whole range of possibilities.
It is not in the least obvious that this is correct. It is not
at all obvious that the laws we observe are 'universal' rather
than dependent on the actual distribution of matter in the
Universe. In this contrary view the local laws of physics which
we measure in the laboratory would depend on the structure of the
Universe as a whole.

§2.2   *The relativity of motion*

The second step in the Copernican theory is to ascribe a
motion to the Earth. This had always been the stumbling block in
the heliocentric theory, since it was argued that the diurnal
rotation of the Earth would lead to absurd results, even if the
annual motion round the Sun were admitted as conceivable.
Copernicus presents the arguments in the first book of De
Revolutionibus:

> If, then, says Ptolemy, the Earth moves at
> least with a diurnal rotation, the result must
> be the reverse of that described above. For
> the motion must be of excessive rapidity, since
> in twenty-four hours it must impart a complete
> rotation to the Earth. Now things rotating
> very rapidly resist cohesion or, if united, are
> apt to disperse, unless firmly held together.
> Ptolemy therefore says the Earth would have
> been dissipated long ago, and (which is the
> height of absurdity) would have destroyed the
> Heavens themselves; and certainly all living
> creatures and other heavy bodies free to move
> could not have remained on its surface, but
> must have been shaken off. Neither could
> falling objects reach their appointed place
> vertically beneath, since in the meantime the
> Earth would have moved swiftly from under them.
> Moreover, clouds and everything in the air
> would continually move westward.
> (I De Revolutionibus, Chapter 8)

To counter this argument, Copernicus contends that the motion of

the Earth is a 'natural' motion and therefore cannot produce
results which are 'against nature',

> Now if one should say that the Earth moves,
> that is as much as to say that the motion is
> natural, not forced; and things which happen
> according to nature produce the opposite
> effects to those due to force. Things sub-
> jected to any force, gradual or sudden, must
> be disintegrated, and cannot long exist. But
> natural processes being adapted to their
> purpose work smoothly. Idle therefore is the
> fear of Ptolemy that the Earth and all thereon
> would be disintegrated by a natural rotation,
> a thing far different from an artificial act.
>   (ibid)

This can be compared to more recent attempts to detect an
absolute velocity of the Earth by local experiments, such as the
Michaelson-Morley experiment (Chap. 7). In all of these cases
the failure to obtain a positive result is embodied in a principle
of relativity. The Copernican principle of relativity, as stated
in the above extract, is that the motion of the Earth cannot be
detected through observations of the local motion of bodies. This
is a mild generalisation of the observed facts, elevated to a
statement of overall principle. It does not explain the fact any
more than the principle of special relativity explains the
Michaelson-Morley experiment. Indeed, Copernicus does not stop
to give the kinematical reason for the absence of retrograde
motion of falling objects and clouds. It is here quite sufficient
that such motions should be required to conform to the general
principle. A further generalisation, in the manner of a later
investigator, would be that an observer in natural motion cannot
by means of any local experiment detect that motion. The generalisation here consists in the reference to *any* observer, rather
than just an earthbound one, and local experiments of *any* type,
rather than merely local kinematics. Such a widening of the
scope of the principle was, of course, beyond the range of the
Copernican theory.
The restriction to local experiments in the Copernican
principle of relativity, or its generalisation, is essential. By
this we confine ourselves to the observation of the motion of
systems shielded from external influences. Merely a glance at
the heavens is sufficient to reveal a relative motion of the Earth,
but this is not a local observation in the sense implied here, so

does not violate the principle.

Copernicus did not really appreciate the essence of Copernican relativity. Certainly he was aware of the relativity of motion:

> A seeming change of place may come of movement
> either of object or of observer, or again of
> unequal movements of the two (for between equal
> and parallel motions no movement is perceptible).
>
> (ibid)

This relativity requires that only relative motion be detectable. Nevertheless, while apparently arguing for the relativity of motion, Copernicus argues at the same time for the reality of absolute motion:

> And why not grant that the diurnal rotation is
> only apparent in the Heavens but real in the
> Earth? It is but as the saying of Aeneas in
> Virgil - 'We sail forth from the harbour, and
> lands and cities retire.' As the ship floats
> along in the calm, all external things seem to
> have the motion that is really that of the
> ship, while those within the ship feel that
> they and all its contents are at rest. (ibid)

If we admit only relative motion then the Ptolemaic objections to the heliocentric theory are groundless. Yet Copernicus could not admit the logic of his own position. Were he to do so, then his own arguments for a heliocentric theory would be groundless too. For the motion of the Earth round the Sun or the Sun round the Earth would represent a mere semantic distinction of no substance.

Nevertheless, there is a distinction between the Copernican and Ptolemaic theories, and we have already given the answer to this paradox. The resolution is that within a purely kinematic treatment of planetary motion carried out on the basis of simplicity of construction, there is a privileged frame with respect to which the laws of motion are stated. This frame of reference differs in the two theories. This is quite compatible with a principle of relativity which states that the preferred frame cannot be found by local experiments. The object of the theory is to find it by global observation of relative motion.

On the other hand, we know that the solution of the problem of planetary motion is not to be found in kinematical theories. We need to turn to dynamics. In a purely kinematical theory it

would be possible to use Kepler's laws to support the Copernican
view, since these give a simple and, at least to a first approximation, a satisfactory account of the motion of the planets.
This is achieved by relinquishing the idea of exactly circular
motion as the only natural periodic trajectory, and allowing the
fundamental paths to be elliptical. However, it would not be
correct to adduce this argument in favour of Copernicus once
Kepler's laws are regarded as deductions from Newtonian theory,
since Newton's is a theory of dynamics. In Newtonian theory it
is still valid to ask, who was right, Ptolemy or Copernicus? The
answer is undoubtedly Copernicus. For, as we shall see, the
Newtonian equations of motion require for their statement a
privileged frame which, in the case of planetary motion, may, to
a first approximation, be taken as a frame in which the Sun is at
rest. It is sometimes contended that in general relativity the
distinction between the Copernican and Ptolemaic systems again
becomes purely semantic. However we shall see that general relativity supports the heliocentric view, but for a rather subtle
reason.

Chapter 3: Newtonian Dynamics

Newton's 'Principia' marks the beginning of modern science in its emphasis on calculation, not explanation. In this it contrasts not only with the Continental followers of Descartes, but also with Newton's own work on optics. 'The Cartesians explained everything and calculated nothing; Newton calculated everything and explained nothing' (Thom 1975). This is the real significance, even if it is not the literal meaning, of Newton's famous 'hypotheses non fingo'. Indeed, Newton's writings on dynamics make more sense when viewed as an attempt to justify the use of a mathematical structure which seems to work. From Newton to the present day the aim of physical science has been the construction of calculable models.

§3.1   *The Law of Inertia*

Dynamics begins with the recognition of 'force' as an entity independent of the motion to which it gives rise. Such a concept was not entirely lacking in Aristotle, where contact forces are permitted to impart unnatural motions; and the process is not completed until the emergence of quantum field theory in which forces are associated with the independent existence of material particles. Nevertheless, it is the recognition that it is the law of force that has to be specified in order to predict motions, together with the mathematical achievement of a technique for working out those motions, that gives rise to the major advances made in Newton's theory.

We can see this already in the First Law, or Law of Inertia, according to which a body subject to no force will remain in its state of motion at constant velocity or at rest. We can rewrite this in analogy with our statement of the corresponding law of Aristotelian dynamics:

> First Law: If a body is in non-uniform motion
> it is necessarily being kept in non-
> uniform motion by a force.

Newton was not unaware of the problem which arises in connection with this law, that of the relativity of motion. Thus a body which is at rest according to one observer has a velocity according to a second observer in relative motion, and a non-uniform velocity according to an observer who is relatively accelerated. The first law, however, singles out a privileged class of bodies whose absolute motion is asserted to be uniform, and it achieves this by identifying those bodies as bodies subject to no forces. Newton implies that the existence of such a class of preferred motions has a dynamical meaning even if it does not have a kinematical one:

> As the order of the parts of time is immutable,
> so also is the order of the parts of space ...
> And so, instead of absolute places and motions,
> we use relative ones, and that without any
> inconvenience in common affairs; but in
> philosophical disquisitions, we ought to
> abstract from our senses and consider things
> themselves, distinct from what are only sensible
> measures of them. For it may be that there is
> no body really at rest to which the places and
> motions of others may be referred.
> (Philosophiae Naturalis Principia Mathematica,
>   Scholium to Definitions)

This idea that the class of uniform motions, the so-called 'inertial' motions, are to be defined dynamically by the absence of forces is reinforced by considering Newton's famous bucket experiment. Here Newton observes that the surface of the water in a rotating bucket will be curved even when there is no relative rotation of the bucket and the water. The curvature demonstrates the existence of forces, and it is this which shows the absolute motion of the system.

Quite independently of whether our interpretation of Newton's meaning is correct, we shall find that it is indeed the case that the force-free motions, the existence of which is asserted by the Law of Inertia, play a fundamental role in the theory. It is by means of these privileged trajectories that we map out the structure of space-time. Newton certainly did not take this next step with much success. For, in constructing a space-time arena for his dynamics he reverts to the idea of a kinematical description of inertial motions. In so doing, not only does he lay himself open to a persistent attack from critics on this point, but, as we shall see, he arrives at a space-time structure which is not strictly consistent with his Law of Inertia. Indeed, Newton constructs the space-time structure appropriate to precisely the Aristotelian Law of Inertia.

We have here an expression of the importance of successful computation. Aristotle's dynamics is completely consistent but does not describe the world correctly. The Newtonian theory is philosophically flawed but is almost always able to produce the right answers. Of course, the flaws can be and have been corrected, but not until after the failure of Newtonian theory to predict correctly had forced on us the new world-pictures first of special relativity and subsequently of the general theory. With the hindsight afforded by these theories it is possible to put classical mechanics on a firm foundation. It turns out that the generalisation of Newtonian dynamics to special and general relativity then appears quite naturally, and that in no way was it required that the historical and logical order should have been different. One might draw a parallel here with quantum theory. Here too we have a highly successful theory with philosophical foundations which to many seem unsatisfactory. Even as apparently extreme a representative of the 'hypotheses non fingo' school as Dirac has, on occasion, voiced disquiet (Dirac, 1973). Which of the two possible morals one might draw from such an historical parallel, however, we leave to the reader's taste.

§3.2   *The Second Law*

Newton's second law relates the magnitude of an imposed force to the acceleration it produces; in symbols,

$$\underset{\sim}{F} = m\underset{\sim}{a} \ .$$

Since we are not here concerned with specific computations, the form of this relation is not important. Note, however, that it is not an identity, and hence it is not a definition of any of the terms in the relation. The mass is a property of the body and the acceleration depends on its motion. The product of these is equated to a force which is supposed to be independent of the body on which it is acting. It is a property of the environment of the body and, perhaps, a 'coupling constant' which measures the strength of interaction of the body and its surroundings independently of its motion. This distinction is crucial to Newtonian theory, for without it any motion is possible. In that case the second law would merely give an alternative description of the non-uniformity of the motion in terms of a force per unit mass.

It is sometimes contended that force is an eliminable concept in Newtonian dynamics. This does not appear to be true. We have already noted that the second law is not a definition of force, so cannot itself be used to eliminate a derived quantity. We may often substitute expressions for forces in terms of relevant fields, for example in considering electrically charged particles in ambient electromagnetic fields. But here 'field' is the derived concept to which 'force' is logically prior. Similarly, the introduction of a Lagrangian or Hamiltonian in certain problems cannot be said to eliminate 'force' in favour of 'energy'.

It is also sometimes suggested that the first law is redundant since it follows from the second. Indeed, the relation between mass times acceleration and force must be such that the acceleration vanishes for zero force: from $\underset{\sim}{F} = m\underset{\sim}{a}$ it follows that $\underset{\sim}{F} = 0$ implies $\underset{\sim}{a} = 0$, and that a body subject to no forces travels with constant velocity. Such an implication correctly expresses the compatibility of the two laws, but completely conceals the significance of the first law. This law of inertia defines for us a privileged class of frames from which the dynamical drama can be observed to unfold according to the Newtonian laws. These frames are those of observers in inertial motion, and are the so-called 'inertial frames'. The first law postulates that such reference frames exist, which the second law does not require since the force need never be zero. And, furthermore, it states that such frames should be identified at the outset since, as we have discussed, such frames are to be used to judge the non-uniformity in the motions of other bodies. Only after the completion of these preliminaries can the dynamical action begin. And for the most part it is only with such preliminaries that this book is concerned.

In order to measure velocities and accelerations we need
to measure time intervals. A uniform velocity according to one
clock will be an unsteady motion according to another clock which
is gaining or losing relative to the first clock. This would not
matter in the study of dynamics if the laws of dynamics were none-
theless the same in both cases, however significant it might be
in other spheres of activity. Unfortunately, it is possible to
set up purely dynamical experiments by subjecting bodies to forces
determined to be the same independently of the clocks, and
measuring their velocities after a given number of ticks of the
clocks. This will distinguish the two clocks. Mathematically,
Newton's laws are not invariant under changes of time coordinate,
apart from a trivial change of origin and of the units of time
measurement. Notice that *a priori* this need not have been the
case. For example, were Newton's second law a definition of
force, it would not be possible to distinguish the two situations.

It follows from this discussion that in order to apply the
dynamical laws, one must identify, physically, a time to which
they refer. To such a parameter Newton gave the name 'absolute
time':

> Absolute, true, and mathematical time, of itself
> and from its own nature, flows equably without
> relation to anything external, and by another
> name is called 'duration'; relative, apparent,
> and common time is some sensible and external
> (whether accurate or unequable) measure of
> duration by means of motion, which is commonly
> used instead of true time, such as an hour, a
> day, a month, a year.   (ibid)

A theory as remarkably successful as Newton's cannot contain terms that are without meaning. Yet the meaning of the time
parameter t and its identification with some physical entity is
elusive. It cannot be the time measured by earthly clocks, since
these do not remain stable over long periods; it cannot be the
time mapped out by the regular orbits of the planets, since, owing
to the perturbations of their orbits by other planets, these are
not precisely regular.

> It may be that there is no such thing as an
> equable motion whereby time may be accurately
> measured. All motions may be accelerated and

retarded, but the flowing of absolute time
is not liable to any chance. The duration
or perseverance of the existence of things
remains the same, whether the motions are
swift or slow, or none at all; and therefore
this duration ought to be distinguished from
what are only sensible measures thereof and
from which we deduce it ...   (ibid)

Newton clearly attempts to resolve the problem by conjuring
up an image of a world existing in some eternal time to be discovered through the application of the exact laws of dynamics to
the given systems, which include both earthly and heavenly
'clocks' with all their imperfections. In this way two quite
separate ideas get mixed up.

First of all, the theory is so successful because there
happen to exist clocks which are sufficiently accurate for the
purpose of verifying Newton's laws. If there were not, we should
not be able to use Newton's laws and we should say they were wrong.
Which of the many possibilities we choose for a fundamental
reference clock is a matter of convention. Hence the existence
of several different reference times - sidereal time, atomic
time ... - and conversion tables relating them. A system based on
my wrist-watch would be in principle possible but quite impracti-
cable. In terms of my wrist-watch time, the frequencies of
spectral lines everywhere in the universe appear to undergo
apparently random fluctuations. Whether any of these conventional
times is compatible with Newton's laws is a matter of experiment.
While experiment appeared to confirm the compatibility it was
possible to regard Newtonian dynamics as the ultimate truth. When
more accurate experiments showed otherwise, it proved more useful
to adhere to the convention of atomic time as the 'absolute time'
of special relativity, and to modify the dynamical laws. In this
way a real, not a hypothetical clock could be used to give meaning
to the laws of motion.

The second concept of absolute time, common to Aristotelian
and Newtonian theory, but not to succeeding theories, involves the
idea of absolute simultaneity. Once the manner of specifying the
time parameter has been chosen, it is assumed that we can speak
meaningfully of the same time at different places. This is forced
on us because all inertial observers are required to use the same
parameter t in the equations of motion, independently of their
location and their velocity. We have already stated that no non-
trivial transformation of the time coordinate is possible at all
if the form of the second law is to be preserved, so it follows

that no transformation of the special type contemplated here is possible. (In fact the second law is $\frac{d^2x^\mu}{ds^2} = F^\mu$ in space-time, and $F^0 = 0$ in *all* frames implies $t = as + b$, where a,b are constants. Hence all times are related by transformation of units and choice of origin.) In space-time terms we again have the existence of a projection π_T from the space-time manifold on to the time axis, dividing the manifold into a stacking together of slices of constant absolute time, each of which represents the whole of space at the same time (Fig. 3.1). This is to be distinguished clearly from the notion of 'absolute space', which arises from an identification of the same place at different times.

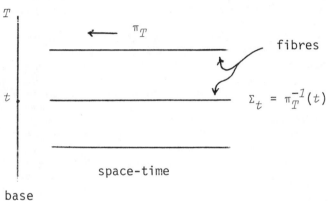

Fig. 3.1. The projection π_T corresponding to absolute time T.

§3.4 *Absolute space*

The projection π_T provides a fibering of the space-time manifold with T, which is just the real line, as 'base' manifold and the constant absolute time sections $\Sigma_t = \pi_T^{-1}(t)$, $t \in T$, as fibres (Fig. 3.1). The existence of an absolute space requires the existence of a projection, π_Σ, from M into some base manifold Σ, which is then the absolute space. As yet we do not have such a structure. That is not to say we do not have any projections π_Σ. Indeed, we have many, since the world-lines of any set of non-intersecting particle paths provides such a projection, the world-lines being the fibres (Fig. 3.2). The point is that at the moment we have no reason for considering any one of these projections as preferred. Thus, kinematically, any observer may con-

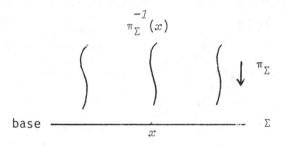

$$\pi_{\Sigma}^{-1}(x)$$

base ——————————————— Σ

x

Fig. 3.2. A projection π_{Σ}.

sider himself to map out the same place at different times no
matter what his motion relative to other bodies. In this way,
any observer can define a projection locally, which he is entitled
to regard as at least on an equal footing with the projection of
any other observer.

Again it is to dynamics that we must turn for additional
structure. Here, as Newton realised, the dynamical laws do not
hold for all observers, but only for a special class of observers,
the inertial observers. Such observers, as laid down by the first
law, have absolutely zero acceleration. Apparently, therefore, we
can distinguish, on dynamical grounds, between absolute motion and
absolute rest. Consequently, underlying absolute motion is absolute space:

> Absolute motion is the translation of a body
> from one absolute place into another, and
> relative motion the translation from one
> relative place into another.   (ibid)

And absolute space is defined by that privileged class of trajectories at absolute rest, the world-lines of absolute immobility:

> Absolute space, in its own nature, without
> relation to any thing external, remains
> always similar and immovable.   (ibid)

In this way, Newton constructs a space-time structure for
his dynamics, which is precisely the structure of Aristotelian
space-time, a produce $\Sigma \times T$ with a privileged vector-field provided by bodies at rest used to define the privileged projection

31

π_{Σ}. Once one has understood the flaw in this argument it is difficult to present it in a way which completely conceals the flaw. Our sleight of hand consists in passing from absolute acceleration, which really *is* required by the dynamical laws, to absolute velocity, which is not, according to Newton's own laws, detectable by dynamical experiment or observation. For, as we shall see more fully later, inertial observers may be distinguished by their lack of acceleration, but they may have any constant velocity. No one amongst the inertial observers moving with constant relative velocity is more privileged than any other. Consequently, there is no privileged projection π_{Σ}, no one definition of the same place at different times, and Newtonian theory does not, in fact, admit an absolute space.

§3.5 *Universal gravity*

As is well known, Galileo did not drop a brick and a feather from the tower of Piza; he was too clever an experimenter for that. It has, however, been shown by more subtle and accurate means that the gravitational attraction of the Earth imparts an acceleration to bodies which is independent of their masses. The force on a body due to gravity depends on the strength of the coupling between the body and the ambient field, and is measured by the passive gravitational mass m_p. The response of the body to a given force depends on the inertial mass of the body m_i, as measured, for example, by the ballistic balance. This is therefore the mass which appears in the expression 'mass times acceleration' in Newton's second law. Applying this law to a body in a gravitational field we have

$$m_p \, g \;=\; m_i \, a \; , \qquad\qquad (3.5.1)$$

where g is a quantity, having the dimensions of acceleration, which measures the strength of the field. Now our experiments tell us that the accelerations of bodies due to gravity at the same point in space and time (or strictly speaking at neighbouring points) are independent of the nature of the bodies. By a suitable choice of units we can take $a = g$, so g is the acceleration due to gravity, and therefore $m_p = m_i$. Experiments establish the equality of the passive and inertial mass to one part in 10^{11}. This is not yet a theory of universal gravity. For while it implies that all bodies are affected equally by a gravitational field if such a field is acting, it does not yet assert its universal action, that is, its infinite range. And while it tells us the acceleration due to gravity is equal for all bodies, it does

32

not yet tell us what that acceleration is, or how it depends on the environment. This information is contained in Newton's famous inverse square law

$$\underset{\sim}{F} = - G \frac{m_1 m_2}{r^2} \frac{\underset{\sim}{r}}{r} \qquad (3.5.2)$$

giving the force between bodies of masses m_1 and m_2 in terms of their separation $\underset{\sim}{r}$ and the universal constant G. Strictly

speaking we should distinguish between active and passive gravitational mass. Thus the force exerted on body *1* by body *2* depends on the product of the active mass of *2* and the passive mass of *1*, and conversely for the action of *1* on *2*. Newton's third law asserts the equality of action and reaction from which it follows that these forces must be equal, hence that the active and passive mass must have a universal ratio common to all bodies, and that by choice of G this ratio may be taken as unity.

Note that the gravitational constant G is dimensional, having the dimensions {mass}$^{-1}${length}^{3}{time}$^{-2}$. As we shall have reason to discuss later (Sect. 11.2), it is not possible to measure dimensional quantities but only their dimensionless ratios, or pure numbers. And indeed, if the only force in nature were gravity it would be impossible to measure G. We could, for example, rescale m, replacing it by $G^{\frac{1}{2}}m$. This quantity then cancels from equation (3.5.1) when we try to measure the force by its dynamical action, there being no other way if there are no other types of forces. It is not possible to measure the mass m independently if only gravitational forces exist. For example the ballistic balance depends on the operation of electrical forces holding matter together. It follows that the role of G in the theory is to measure the ratio of gravitational to other forces. For example, a comparison of the ratio of the gravitational to electrical forces acting between two electrons is given by the dimensionless number $G m_e^2 / e^2 \sim 10^{-38}$, where m_e is the electron mass and e its charge. Why it is that gravity is so weak compared to other forces is a fascinating and complete mystery.

An important feature of the Newtonian law is its 'linearity'. That is, the net force on a given body due to a distribution of masses is the vector sum of the forces that each mass would produce if it alone were acting. Not all theories have this property, so it is not obvious in any way. For example, in general relativity, the gravitational affect of a given mass depends to some extent on what other masses are present and where they are.

Rather than working with forces it proves useful to work with the related concept of potential. We let $\phi(\underset{\sim}{r})$, the potential at $\underset{\sim}{r}$ corresponding to a given fixed distribution of gravitating

33

masses, be the work done on a unit mass in bringing it from infinity to $\underset{\sim}{r}$. Then, the force on a mass m at $\underset{\sim}{r}$ is $\underset{\sim}{F} = - m \underset{\sim}{\nabla} \phi$.
From the superposition law for the forces follows a superposition law for the potential. Thus, for a distribution of mass with density $\rho(\underset{\sim}{r})$, we have

$$\phi(\underset{\sim}{r}) \quad = \quad - \int \frac{G \, \rho(\underset{\sim}{r}')}{|r - r'|} \, d^3r' \qquad . \qquad (3.5.3)$$

This is the integral form of Poisson's equation,

$$\nabla^2 \phi \quad = \quad - 4\pi G \rho \qquad , \qquad (3.5.4)$$

and both versions are exactly equivalent to Newton's law together with the principle of linear superposition. These remarks will be important in our discussion of general relativity and 'Mach's Principle' in Chapter 12.

§3.6 *Dynamics in a gravitational field*

By providing a general expression for the gravitational force, the Newtonian theory of gravity gives rise to a complete description of the motion of bodies under their mutual gravitational influence. In principle, all that is required is the solution of the differential equations, provided by the second law, for the coordinates of the particles as functions of time in a given inertial frame. To obtain a geometrical picture of this situation one might represent the gravitational field in terms of 'lines of force' in analogy with the customary representation of a static magnetic field. The tangent vector to a gravitational line of force at a point would be the direction in which a body at rest at that point would start to move (provided the body is sufficiently small so as not to sensibly affect the gravitational field we are representing).

However, in contrast to the Copernican picture, for example, this representation in terms of lines of force, or their tangent vector fields, does not provide an immediate solution of the problem, since at each point it yields directly only the initial space-time trajectory of a body at rest, not of one in motion. The same vector field is associated with many space-time trajectories depending on the initial velocity of the body.

The problem can be approached from a slightly different point of view. It is known that the theory of statics is governed by a single principle, the principle of virtual work, which states in essence that an equilibrium situation is one in which the

34

potential energy of the system is a minimum. One can ask whether there is in the theory of dynamics a function which will play a role analogous to that of the potential energy in statics. The answer to this was found to be affirmative by Lagrange and the appropriate function is called the 'Lagrangian', denoted by L. The Lagrangian is a (real-valued) function of the coordinates and velocities of the particles in the system. It is therefore defined on the space of all coordinates and tangent vectors called the 'tangent bundle' to the configuration space Σ (Fig. 3.3).

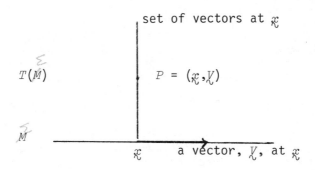

Fig. 3.3.   The tangent bundle $T(\Sigma)$ to a manifold Σ. Each point P of $T(\Sigma)$ represents a point in Σ and a vector at that point.

(As stated, this seems to require the existence of an absolute space Σ; we simply note that in fact all that is required is the identification of inertial frames.) The function L gives rise to a certain association of a number with any trajectory between two points: the path for which this quantity is a minimum is the solution of the dynamical problem.

   Now pass from this global statement to a more convenient local formulation by passing from L to a new but related function, H, defined on the space of coordinates and momenta. H is called the Hamiltonian and the space on which it is defined is the 'cotangent bundle' to the configuration space Σ (also called 'phase space'). In simple situations, at least, the Hamiltonian is the total energy of the system, not merely the potential energy, with the proviso that it is expressed as a function of coordinates and momenta, not velocities. Now H gives rise to a vector field in phase space: if we imagine H to define a surface by plotting its value at any point as a height above that point, then this vector field ∇H at a point lies in the direction of maximum slope of the surface there, and has a magnitude equal to the gradient (Fig. 3.4). This vector field, as usual, corresponds to a system

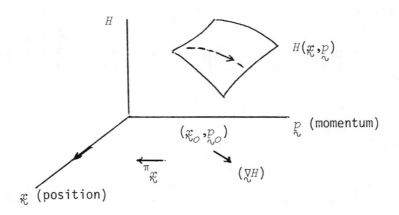

Fig. 3.4.   The description of motion in terms of vector fields (∇H) in phase space ($\underset{\sim}{x}, \underset{\sim}{p}$).

of trajectories, which are just the possible particle motions. If we project these trajectories from phase space into configuration space Σ, by considering only the coordinates and not the momenta, we obtain the particle trajectories in space, and hence the paths through space-time.

Thus, while the purely kinematical Aristotelian theory specifies the motion in terms of vector fields on M (space-time), the dynamical theory of Newton is equivalent to a specification of the motion by means of vector fields on the co-tangent bundle to M or, equivalently, on phase space. Note that the motion so described is not related in any simple way to the vector field on M determined by the gradient of the gravitational potential, $\nabla \phi$. This description conceals profound computational difficulties if one wishes to study the motion of a system of N bodies, the gravitational fields of none of which can be neglected - the N-body problem. For in order to determine the appropriate H for any one body we need to know the potential energy of that body, hence the positions of all the other bodies. But this is precisely what we are trying to determine in the first place.

Chapter 4: Critiques of Newtonian Dynamics

The impressive successes of the Newtonian theory in the
development of our understanding of matter in motion rivalled even
those of Euclidean geometry in its description of the measurement
of space. Thus, if the latter can be established on the basis of
a set of *a priori* acceptable axioms, could it not be that the
truth of Newton's theory too could be equally firmly established?
Against this view were set the voices of those for whom the Newtonian Absolutes were a source of grave intellectual disquiet.
To a certain extent, this disquiet arose from a need to unify the
scientific and theological pictures of the world in a way which is
now generally considered to be unnecessary. Nevertheless, with
some of its more lucid expositors, the unease was based on valid
scientific intuition, which, first raised in the contemporary
writings of Leibniz, and Berkeley, and later in the positivisti-
cally oriented works of Mach, led to Einstein's development of
general relativity.

§4.1 *The Leibniz-Clarke controversy*

There are two main lines of argument running through the
more closely scientific aspects of the correspondence between
Leibniz and Newton's disciple Clarke. The first of these, while
not directly relevant to the nature of the space-time structure of

37

Newtonian theory, will nevertheless be of relevance to us later. This is the problem of the stability of planetary orbits. The second line of discussion concerns the meaning of absolute space and time.

The first problem arose because it was believed by Newton and his contemporaries, on the basis of their calculations according to Newtonian theory, that the orbits of the planets were secularly unstable. This means that the planets would not remain in their approximately elliptical orbits, but that a given planet, the Earth say, would be gravitationally attracted by other planets causing a departure of the orbit from an exact ellipse. This discrepancy would built up without limit as time went on. In this way perturbations would grow and the system would break up. The conclusion was that the regular clockwork of the heavens was running down.

This is not an immediate physical problem if you believe, as Newton did, that the world was created a finite time ago, provided that this beginning is too recent for the perturbations to have built up by now to values larger than those observed. It becomes a serious problem if one insists, as again by implication Newton presumably wanted to, that the planetary system would be unchanging for ever in the future. For in that case the clockwork needs to be put right from time to time if the laws of dynamics are to be compatible with a stable system. The secular instability calculations that were the point of discussion between Leibniz and Clarke are now known to be erroneous. In fact, the question of the ultimate stability of the solar system has not yet been settled, but there is no conflict now between observation and theory.

The issue can be raised in more general contemporary terms: do the equations of the theory, together with the positions and velocities of all the particles in the universe at some time, uniquely determine the future dynamical evolution of the system? In the consistent modern version of Newtonian theory the answer, surprisingly enough, is that in general they do not (see Ellis, 1971). For the equations of Newtonian theory applied to the Universe as a whole allow the possibility that certain gravitational fields can exist which are not caused by material particles, and, furthermore, these fields can be spontaneously created. Whether such fields are to be regarded as physically reasonable is a quite different question, as is whether they could indeed be of such a type as to ensure the stability of the solar system. The issue is of relevance since, as we shall see, it is quite easy to construct models in general relatively in which the future evolution of the system cannot be predicted, and it may well be that some of these are physically relevant. If one were to ask here, what does this imply? two answers can be envisaged: either general relativity is still incomplete, or else the world really may have a time limit. Should we have been able to submit the matter to Leibniz and Clarke then we might expect their answers to be along the following lines. Leibniz would contend that the theory of general relativity cannot be true, since both in principle and in fact "God has foreseen everything. He has provided a remedy for everything beforehand" (Second Letter of Leibniz, no. 8). Clarke would be of the opinion that general relativity, that marvellous generalisation of Newton's principles, supplies one more argument for the thesis that in certain circumstances God must intervene to sustain the world in its existence.

Leibniz's second line of attack on Newtonian dynamics concerned a criticism of the concepts of absolute space and time. Leibniz believed that space and time exist only as relations between matter, not as independent 'absolute' entities. A fair analogy is the existence of the family-tree which exists as an expression of biological relationships between its members. Leibniz writes:

> I will here show how men come to form to themselves
> the notion of space. They consider that many things
> exist at once, and they observe in them a certain
> order of coexistence, according to which the
> relation of one thing to another is more or less
> simple. This order is their situation or distance.
> When it happens that one of those coexistent things
> changes its relation to a multitude of others which
> do not change their relations among themselves, and
> that another thing, newly come, acquires the same
> relation to the others as the former had, we then
> say it is come into the place of the former; and
> this change we call a motion ... And that which
> comprehends all those places is called space.
> Which shows that in order to have an idea of place,
> and consequently of space, it is sufficient to
> consider these relations and the rules of their
> changes, without needing to fancy any absolute
> reality out of the things whose situation we
> consider.   (Fifth Letter of Leibniz, no. 47)

This bears a striking resemblance to Einstein's discussion:

We come now to our views on space. An
important thing here is to take notice
of the relation of experience to our
concepts ... By means of simple changes
in their position, we can put two bodies
together. Theorems on congruences, which
are fundamental to geometry, are linked
with laws governing such changes of position.
To grasp the concept of space, the following
remarks seem essential: We can form new
bodies putting bodies B, C ... next to A;
we then say that we have extended body A.
Extending sufficiently any body A, we can
make it to reach any other body X. The set
of all such extensions of body A we shall
call the space of body A. It is accordingly
a true assertion, that all bodies exist in
the space of a (freely chosen) body A. In
this manner, we do not allow ourselves to
speak of "space" - abstractly - but of the
space relative to a body A.

(Einstein, 1950)

We see that Einstein and Leibniz are arguing that space is
a property of things and not itself a 'thing'. Similar remarks
can be made about time. Thus space is 'an order of co-existence
as time is an order of successions' (Third Letter of Leibniz, no.
4). And this is in direct conflict with Newton who contended that

... the instants of time are objects
analogous to events, but different from
them. As such they do have properties,
but are not the properties of anything;
in particular, they are not the properties
of events. In consequence ... time exists
independently of the material world.

(Augustynek 1970 pp.20-21)

In support of this view, and against the relational theory
Clarke could argue that 'space and time are quantities, which
situation and order are not' (Clarke's Third Reply, no. 4). In
the case of the analogy of a discrete genealogical network the
naming of the relation (e.g. 'father', 'second cousin twice removed', and so on) can itself be used to specify at least some
sort of quantitative measure of the closeness of the relationship -

which modern genetics could no doubt make precise. In the setting of a continuous space-time the order relation, which, as we have seen in Chapter 1, defines a topology, is quite separate from the metric properties. Einstein's argument shows only how space-time relations can, in principle, be established, not how space is derived from matter, that is, not how the possibility of those relations is provided by the material condition of the world.

Furthermore, the relational theory gives no account of inertia. In his third reply to Leibniz, Clarke says

> If space were nothing but the order of things
> coexisting, it would follow that if God should
> remove in a straight line the whole material
> world entire, with any swiftness whatsoever,
> yet it would always continue in the same place,
> and that nothing would receive any shock upon
> the most sudden stopping of that motion.
> (Clarke's Third Reply, no. 4)

Followers of Leibniz would doubtless argue that this is just as it should be, since the motion of the world as a whole can have no meaning in the relational theory. But the argument can be applied to any single body, since a chosen body can be given an arbitrary motion by viewing it from the appropriate frame of reference. However convincing the kinematic argument might be, the true ground for the contest must be dynamics. And here, of course, Newton had a theory which worked and Leibniz did not.

However, the Leibnizian relational viewpoint is not wrong *per se*, and one can legitimately ask how it fares in general relativity. The most straightforward way of implementing the relational theory of space is to make space a material plenum, as apparently first proposed by Aristotle's pupil Theophrastrus (Jammer, 1960). The properties of space-time then become trivially dependent on matter, and space-time can be regarded as the property of spatio-temporal extension of matter. In general relativity, Einstein's idea seems to have been in part to replace the material plenum by the 'metric field', that is, an entity existing at each point of space-time which defines the local geometrical properties of space-time. As Einstein was quick to realise, this is not sufficient, at least in the form that it takes in general relativity. For it turns out that space, far away from matter, has all the properties with regard to inertia attributed to Newtonian absolute space. We shall return to this later in more detail (Chapter 9). One might note here that the requirement of the Leibnizian viewpoint is even stronger. Leibniz space-time simply cannot exist if

41

there is no matter. And similarly, as noted earlier by St. August-
ine, the concept of time cannot appear if there is no motion.

As yet there is no physical theory embodying these ideas
and the Leibnizian view remains only an attractive programme. In
general relativity, as in Newtonian theory, the space-time is a
given manifold, the structure of which is to be constructed
according to the material possibilities. For the present, space-
time is a 'thing'; and some of its properties are determined or
modified by matter.

§4.2 *Relativity and Phenomenalism*

> Time, place, and motions, taken in particular
> or concrete, are what everybody knows; but,
> having passed through the hand of a metaphysician,
> they become too abstract and fine to be
> apprehended by men of ordinary sense.
> (Berkeley, On the Principles of Human Knowledge,I,XCVII)

Berkeley's criticism of Newtonian mechanics arises from his
general attack on the notion of 'abstract ideas', as propounded in
particular by Locke. For Berkeley physical objects are bundles or
complexes of directly perceived qualities, a standpoint which can
be characterised as 'phenomenalism'. The idea of an abstract
reality behind these qualities is absurd; for, by definition, such
a reality cannot be perceived and is consequently unknowable.
Berkeley's philosophy is therefore, he believes, an appeal to
common sense, as reflected in the above quotation from The Principles
of Human Knowledge.

One of the applications of this common-sense point of view
is to the philosophy of Newtonian mechanics. 'If we sound our own
conceptions', writes Berkeley, 'I believe we may find all the absolute motion we can frame an idea of, to be at bottom no other
than relative motion' (The Principles, I, CXIV). This follows,
because to conceive motion we must necessarily imagine at least
two bodies the distance between which is changing, so that the
idea of motion 'doth necessarily include relation' (ibid. I, CXIII).
In principle one could think of bodies moving against a labelled
grid, or backcloth, but we have no perceptual awareness of such a
grid. Absolute motion is an 'abstract idea', since it cannot be
perceived by the senses, and is therefore meaningless.

In fact, this lack of meaning becomes completely clear if
we consider uniform motion. For 'we cannot know ... whether the
whole frame of things is at rest, or is moved uniformly' (De Motu,
65).

42

These arguments against absolute motion are purely kinematical, and it was clear to Newton himself that from a kinematical
viewpoint all motion is relative. This indeed was precisely the
source of his difficulty. For, from the point of view of dynamics
there appear to exist privileged observers, namely the inertial
observers, and it is to explain the existence of these dynamically
'absolute' reference frames, from which motion appears dynamically
absolute, that Newton found it necessary to introduce the backcloth
of absolute space. In non-inertial frames 'fictitious' forces have
to be introduced which owe their existence to no apparent material
cause. The problem of computing dynamical results is solved once
a suitable reference frame for the application of the laws is
specified, that is, once the force-free, straight-line motions
postulated by the first law are identified to sufficient accuracy.
This was first solved intuitively by Galileo in his theory of the
tides, where he took the reference frame tied to the distant,
apparently fixed, stars as a fundamental inertial frame. In
practice Newton accepted this for the purpose of planetary dynamics, although he could legitimately doubt whether it was in fact
true. This is because a universe is conceivable in which the fixed
stars are not at rest and in which the inertial frames are identified on other grounds. Indeed, if we *define* a *kinematic* inertial
frame as one in which the distant stars appear fixed, and a
dynamical inertial frame as one in which Newton's laws hold without the introduction of fictitious forces, then there is nothing
in these laws themselves which guarantees or requires the agreement of the two frames. From the Newtonian point of view, this is
merely an observed fact at a certain level of accuracy in the
universe in which we happen to live. There are various experimental
tests of this agreement, the simplest (and least accurate) of which
is the Foucault pendulum. There is nothing in Newton's laws to
explain why this pendulum maintains its oscillations in a plane
fixed relative to the stars, rather than with respect to some other
frame. On the contrary, this is taken as an observational input
into the theory, which can in principle be modified if more accurate
observation should so require. Consequently, since the laws hold
for some frame, this frame is designated as absolute space, or,
more correctly, since the laws hold for a class of frames, this
class can be called 'absolute'.

We can see from these arguments that no purely kinematical
solution of the problem is possible if we wish to stay within the
context of Newtonian theory. Kinematically relative motion is
motion relative to an arbitrary standard of rest; to eliminate
absolute motion one does not need the whole Universe but only four
bodies and these can be chosen at random. In Newton's bucket ex-

43

periment the bucket is chosen as the arbitrary standard to which
the motion of the water is referred. The dynamical affect of a
curved surface to the water in the bucket occurs when there is
zero relative rotation between the bucket and the water and not
when the relative rotation is a maximum. This observation
correctly shows that an arbitrary standard of rest, hence the use
of truly relative motion, is inadequate for Newtonian dynamics.

Berkeley tries to solve this problem by making a distinction
between real and apparent motion, or, more strictly in accordance
with what he says, between the true and false attribution of motion
to bodies. The idea, presumably, is that real motion should be
motion relative to the Universe as a whole, hence that only true
accelerations are to be admitted in Newton's laws:

> for ... the purpose of the philosophers of
> mechanics ... it suffices to replace their
> 'absolute space' by a relative space determined
> by the heaven of the fixed stars ... Motion
> and rest defined by this relative space can be
> conveniently used instead of absolutes.
>   (De Motu, 64)

Furthermore, according to Berkeley, true motion (by which he presumably means true non-uniform motion) is distinguished by the
application of forces to bodies. Thus, by local observation we
are supposed to be able to discover if a body is really moving
(accelerating), or if we are mistakenly attributing a motion by
observing it from a moving frame. This would be at least meaningful if all real forces were immediately knowable apart from the
motion they produce. But, 'no force is ... known or measured
otherwise than by its effect' (De Motu, 10). For Berkeley, forces
are abstract ideas, or 'occult qualities', since they are not
immediately perceived, hence cannot be known. Mechanistic explanation in terms of forces is acceptable as a mathematical hypothesis which allows the regularities in Nature to be brought out,
but forces cannot be held to be the causes of the motions described
in terms of them. It follows from this that to describe gravity,
for example, as an inherent attractive power of matter, as if this
were the cause of gravitational attraction, is meaningless. For
since this inherent essence of ponderable matter is not immediately
perceived in an independent way, to ascribe accelerations to
gravitational forces is to name an effect, not to specify a cause.

This causes certain difficulties for the distinction between
true and apparent motion. For suppose I wish to choose the Earth
rather than the stars as the standard of rest. In order to success-

fully describe the motion of the planets all I need do is to introduce forces according to which the planets follow their prescribed paths. Admittedly, these forces will be somewhat complicated, being the sum of the inverse square force and the inertial forces introduced by referring motion to a rotating Earth. Nevertheless, in this system the stars are truly in motion, and the Earth is truly at rest.

While, therefore, Berkeley's suggestions leave the Newtonian problem unresolved, he does come close to stating a valid objection to Newtonian absolute space (in addition to the unknowability of uniform translation). Berkeley believes the gravitational force not to be universal ('as in the perpendicular growth of plants'), and hence to be invoked *ad hoc* to explain those motions which it just happens to predict correctly. Were this the case, that is, were the gravitational attraction to be granted to the planets but not the stars, and stones but not air, then the idea of gravity as a property of matter being the cause of certain motion would be patently meaningless. Now precisely this argument can be used against absolute space. For in asserting that absolute space is the cause of inertial forces and in maintaining that absolute space is known *only* through the existence of inertial forces, Newton is not asserting a cause at all, but merely naming an effect. To see this clearly it is only necessary to replace 'absolute space' by 'the body alpha'. Acceleration relative to alpha produces inertial forces and this explains the existence of inertial forces, whereas the existence of inertial forces demonstrates the existence of alpha. Only if astronomers or archaeologists were to find some independent evidence for the existence of alpha would its inertia producing qualities be a scientific hypothesis.

With the success of the theory of universal gravity and Newtonian dynamics, the essentialist doctrine prevailed at least until the criticisms of Mach revived the phenomenalist viewpoints (Popper, 1954). The argument against pure phenomenalism is that some theoretical construction is necessary even for the interpretation of elementary, directly perceived, sense data. Once this is admitted our limited ability to collect data directly can be extended by theoretical construction to, for example, other parts of the electromagnetic spectrum, since we might have evolved with X-ray eyes, or by observations of, say, planetary and stellar dynamics. This latter shows that it is part of the essence of matter to gravitate, just as it is part of its essence to have extension or colour.

This conclusion, however arrived at, is the essentialist viewpoint. Its limitations are apparent from the fact that we know matter not to be essentially coloured, since, for example, a free

45

electron at rest is not coloured. The essentialist view requires at least that we should be in possession of the ultimate and true theory of Nature. However, the discovery of such a theory, and with it the essence of the ultimate building bricks of the Universe, is a scientific task which remains uncompleted - as indeed does the determination of whether such a picture is possible.

Nevertheless, it is of interest to note that the current quantum field theoretical description gives support to the reality of forces. For example, contact forces originate from the electric force between electrically charged bodies (at a microscopic level). The static electric force is known to be simply related to the quantum of electromagnetic radiation, the photon. The force is therefore as real - or as hypothetical - as light, since it is just another aspect of the same theoretical construction. It is ironic that the failure, so far, of attempts to create a satisfactory theory of gravity subject to quantum rules means that we cannot make a similar assertion for the gravitational force.

§4.3 *The space-time of pre-relativistic kinematics*

Leibniz and Berkeley agree on two points, namely that without matter there is no space-time, and that it should be possible to use any body as a reference frame for the statement of dynamical laws, whatever the state of translational or rotational motion of the body. We do not know how to satisfy the former requirement. We have seen that the latter describes the space-time structure appropriate to kinematics, but has not resulted in a successful dynamical theory.

This kinematical structure can be defined in the following way. Leibniz and Berkeley, in spite of their declarations, require an 'absolute time' since they need to consider the co-existence of events. Thus there is the familiar projection π_T from the space-time manifold on to the one-dimensional time manifold. Since time is merely the order of events there is no metric imposed on this time manifold, i.e. no function for expressing time-differences numerically is given *a priori* but only the order of time-instances. There is no projection π_Σ on to an absolute space since this would define preferred motions, namely the world-lines $\pi_\Sigma^{-1}(x)$ with x a point in Σ. This means there is no preferred way of identifying the 'same' point in the surfaces t = constant as t varies. Any points that can be joined by smooth curves can be taken to be the same point at different times, since the curve can be taken to represent the motion of a body which can then be chosen as the reference body of an observer.

The surfaces t = constant are supposed to have a Euclidean geometry, since at any given instant the theorem of Pythagoras is supposed to apply. In particular, it is meaningful to talk of vectors being parallel in $\pi^{-1}(t) = \Sigma(t)$. In contrast, since a reference body is allowed to rotate arbitrarily it is not meaningful to talk of the same direction at different times. This will depend on the arbitrary relative rotation of the reference body. It follows that vectors in $\Sigma(t)$ at different times cannot be said to be parallel, since such parallel vectors would define for us the same direction at different times. Furthermore, vectors not in $\Sigma(t)$ cannot be meaningfully parallel since any two vectors could be parallel for a suitably chosen observer. This is important, because it is precisely the possibility of identifying parallel vectors that distinguishes the space-time of classical dynamics from that of kinematics.

In order to discuss the requirements to be satisfied by a dynamical theory constructed on the basis of the structure provided by the space-time of non-relativistic kinematics, it is useful to discuss this structure from a different point of view. This will also be of help later.

In our discussion of Aristotelian theory we considered the possible alternative choices of time consisting of a transformation of origin $t \to t + t_o$ and of scale $t \to kt$. With the kinematical structure we are now considering, we can be much more general and admit any transformation $t \to f(t)$, provided only that the function f is smooth and monotonic (that is, $f(t_2) > f(t_1)$ if and only if $t_2 > t_1$, so time order is preserved). This would correspond to the requirement that the dynamical laws should be the same whatever clock we might choose to measure time. We can extend the discussion further and ask what choice we have in the labels we attach to points of space. In the case of Aristotelian space-time we have a choice of spatial unit of measurement $\underset{\sim}{x} \to k\underset{\sim}{x}$, and of spatial origin $\underset{\sim}{x} \to \underset{\sim}{x} + \underset{\sim}{a}$ where $\underset{\sim}{a}$ = constant. We imagine, for example, the space Σ to be slid over itself through a displacement $\underset{\sim}{a}$ and ask what difference this makes. Strictly speaking we are not allowed to do this in Aristotle's original picture since the Earth is the centre of a spherical universe, but in our slightly modified model in which Σ is simply R^3, the transformation is allowed, and we find it makes no difference at all: we cannot tell from the results of physical experiment or observation whether or not this transformation has been performed. There is one further transformation that leaves the physical nature of the system unchanged. This is a rotation of the system in space about the origin. The space spection Σ does not have a built in preferred direction at the origin. In Aristotelian space there is a preferred direction at any other point, namely that pointing

47

towards the Earth at the origin, so strictly speaking this rotation cannot be performed at any point other than the origin. This is a difference between Aristotelian space-time and Newton's picture. In the latter there is no preferred direction built in to the space-time structure, and if the Universe were to be rotated through some angle, about any point in space say, by changing the orientation of the laboratory to a new, fixed orientation, this would go undetected (Fig. 4.1).

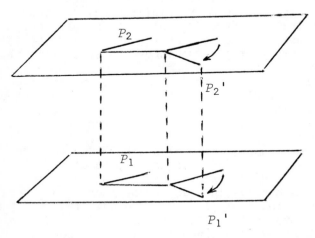

Fig. 4.1.   A 'rigid' translation and rotation of space in
            Aristotelian space-time: points such as P_1, P_2 are
            moved to P_1', P_2'.

If we add up the parameters needed to define these transformations we find one for spatial dilation, three for translation, three for rotation (two to specify the direction of the axis of rotation and one to specify the angle of rotation).   In addition there is one for time dilation and one for time translation making a total of nine real parameters.

The kinematical structure imposed by Leibniz and Berkeley is quite different (Fig. 4.2). We are allowed to slide the surfaces $\Sigma(t)$ over themselves in a different way at each time, so we can make time dependent translations $\underset{\sim}{x} \rightarrow \underset{\sim}{x} + \underset{\sim}{a}(t)$.   In this way any world-line which was initially curved can be made to appear straight. Thus there are no kinematically privileged motions.   In addition we can rotate the spaces $\Sigma(t)$ through arbitrary angles at different times, provided we do this smoothly, and this is described by a transformation involving three functions

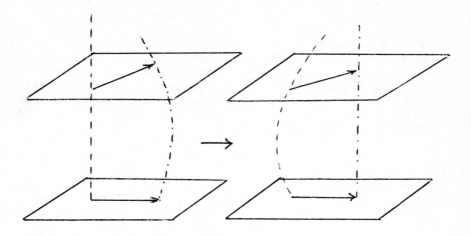

Fig. 4.2. A 'non-rigid' rotation of space by different amounts
at different times in Leibniz-Berkeley space-time.

of time; two to specify the axis of rotation and one to specify
the angle. Therefore the group of allowed transformations in
the space-time of kinematics is described by one parameter dis-
cribing the change of spatial scale and seven functions of time,
if we include the transformation $t \to f(t)$.
        A dynamical theory satisfying the criticisms of Leibniz
and Berkeley would have to consist of laws of motion which are
unchanged in form under these transformations (see Barbour and
Bertolli, 1977). For otherwise the dynamical laws would impose
additional structure. Indeed it is precisely because Newton's
laws are not invariant under this large class of transformations
that they impose additional structure on space-time, and cannot
be considered to be compatible with Berkeley's criticisms as he
appears to think. This is an important point. Notice we are not
saying that the laws of Newtonian mechanics cannot be stated in a
frame of reference with arbitrary acceleration and rotation. Indeed, this can be achieved by performing the appropriate time
dependent transformation to this frame from an inertial frame in
which the form of the laws is known. The point is that the form
of the laws is different in the non-inertial frame, and from this
the non-privileged nature of the frame can be deduced.
        The transformations we have described are a little more
general than is strictly necessary. The existence of an
absolute time means physically that there must exist an absolute

49

clock of which any observer must be aware. It might for example be taken to be defined by the rotation of the Earth about the Sun, or the rotation of the Earth on its axis, etc. Alternatively, it might be determined by the sun-spot cycle or the first cuckoo in Spring. The latter vary monotonically but not regularly with respect to the former. There appears to be little to be gained by such generality and it would seem simpler to state the dynamical laws in terms of an absolute clock time t, it being part of the statement of the theory that the existence of such a clock within sufficient accuracy is to be determined empirically. This corresponds to the procedure in Newtonian theory. It might be objected that all that is required is that the theory predict correlations between observable dynamical phenomena, and that the existence of the absolute clock is redundant. This is true, but it is conventional to designate one dynamical system as a clock and to refer others to this, and it seems unnecessary to break with this convention. The assignment of a clock keeping accurate absolute time corresponds to the selection of a system operating according to purely dynamical laws in a known way. Thus the inclusion of the possibility of the transformation $t \to f(t)$ for arbitrary f adds nothing of significance. It is still necessary to assert the possibility of the 'affine' transformation $t \to kt + t_o$ since this expresses the absence from the space-time structure imposed by classical dynamics of any ultimate unit of time or of an origin for time.

§4.4   *Mach's science of mechanics*

It took some two hundred years for criticisms of Newtonian dynamics to be considered seriously, during which time the literal 'essentialist' interpretation of Newtonian theory prevailed. This was not unreasonable while there was no reason to doubt that the theory was true - and it was difficult to really doubt its validity on the basis of the single observable anomaly in the motion of the planet Mercury. More than this, Newtonian dynamics provided the key to the solution of a continuous stream of interesting problems. One can draw a modern parallel with quantum mechanics. This too has been subject to trenchant criticism which has left it essentially unchanged. For quantum theory appears to be the correct approach to a myriad interesting problems in the atomic domain. Even when this ceases to be so, it is safe to predict that changes will be brought about to satisfy primarily experimental rather than philosophical needs.

Perhaps the most influential critique of Newtonian theory is that of Mach, given in most detail in his 'Science of Mechanics'.

To an extent the work acquires a certain amount of reflected
glory from the fact that Einstein read it was was deeply influenced by it. It has also been argued that Mach was almost
entirely anticipated by Berkeley (Popper 1954; Whitrow 1954).
However, it is idle to speculate whether Einstein would have
read Berkeley had Mach not existed (although his interest in Kant
suggests he might have done so) or indeed whether Mach had read
Berkeley. Furthermore, we hope to show that in important detail
Mach's viewpoint on Newtonian dynamics differs significantly from
that of Berkeley.
    Certainly Mach, like Berkeley, is a supporter of the
phenomenalist creed. Consider, for example, the following passage
of Berkeley from Siris on the role of the natural sciences:

> their province [is] only to discover the laws
> of nature, that is, the general rules and
> methods of motions; and to account for
> particular phenomena by reducing them under,
> or showing their conformity to, such general
> rules.   (Siris, III, 231-232)

And compare this with Mach, in the Analysis of Sensations:

> ... all that is valuable to us is the discovery
> of functional relations, and that what we want
> to know is merely the dependence of experiences
> on one another. It then becomes obvious that
> the reference to unknown fundamental variables
> which are not given (things-in-themselves) is
> purely fictitious and superfluous.
>   (The Analysis of Sensations, p.35)

    Both Mach and Berkeley take the task of science to be to
relate the appearances. For Mach the appearances are everything:
'Bodies do not produce sensations but complexes of elements
(complexes of sensations) make up bodies' (ibid. p.29). And
there is nothing more to discuss. In contrast, for Berkeley
there remains a vast field for metaphysical speculations, since
behind the appearances there is the ultimate reality of God.
    It is easy to criticise phenomenalism as a complete theory
of science, and we have already done so in Section 4.2. In a
sense it is too easy: after all, Mach did not believe in the
existence of atoms, and it is certainly with statements about
the existence of things that phenomenalism has most difficulty.
However, to take a modern parallel, certain features of elementary

particle physics can be explained by postulating the existence of particles called quarks. So far, no-one has found a quark, so they have roughly the same status as atoms had at the end of the nineteenth century. Yet the phenomenalist philosophy does not stand or fall by the discovery of isolated quarks. The problem is not entirely trivial. Nevertheless, as Einstein was to point out as he turned away from Machian philosophy in his later years, there are no *a priori* 'appearances'; only the theory tells you what can be observed.

In analysing Newtonian mechanics from his philosophical viewpoint, Mach repeats the now familiar criticism that on kinematical grounds absolute motion is quite meaningless, since only the motion of a body relative to other bodies can be defined. One can therefore attempt to construct a dynamical theory on the basis of the kinematical structure of space-time discussed in Section 4.3, or one can add additional structure to the space-time. In practice the latter approach proved more useful and amounted to the identification of privileged inertial frames. Mach is concerned with the empirical aspects of this identification. He points out that Galileo chose the Earth as providing a fixed reference frame; that for Newton there was no doubt that the Earth rotated but that, with some reservations, the frame of the fixed stars could serve for the purpose of computing planetary motions. For future astronomers, able to determine with great accuracy the relative motions of stars, and able to see deeper into space to take into account more stars, a correction to the inertial frame will become calculable. This has indeed occurred and it is now possible to discuss the rotation of our Galaxy, containing Mach's 'fixed' stars, with respect to other clusters of galaxies.

Having established a kinematical inertial frame from the observed motion of the stars, then, according to Mach, it is the task of dynamics to relate the appearances. Unfortunately, however successful it may be on a practical level, the frame of the fixed stars does not yet in any way replace absolute space in Newtonian theory, at least as far as absolute acceleration is concerned. Mach was indeed aware of this, and we have already seen that any purely kinematical attempts to abolish absolute space are bound to fail. If the Earth rotates merely relative to the stars then equally the stars rotate relative to the Earth and it is meaningless to distinguish the two cases. Yet dynamically the two cases are apparently distinguished: a rotating Earth is flattened at the poles, a non-rotating one is not.

Mach tried to argue that it is this latter distinction that is in fact meaningless. For we do not observe the world twice, once with an Earth at rest and once in rotation. It is therefore nonsense to pretend one knows what would happen if the Earth were at rest and the stars in rotation as if this were a different case from that of a rotating Earth in a fixed Universe. Only the one case is presented to us, so Newton's Laws must apply to just that one case, which we can interpret in any way that does not conflict with experience.

This is an ingenious, - or to some tastes a curious, - argument (Russell 1903, §469). It goes wrong in attempting to apply Newton's Laws in a purely kinematical setting. If it is true that only the one case of relative motion is presented to us, it follows that Newton's Laws cannot be valid, at least in their present form. Conversely, if Newton's Laws are valid then the one case that is presented to us is the absolute rotation of the Earth relative to a dynamically determined inertial frame. This can be seen quite clearly by considering a situation in which the light of the stars has been turned out, or, at a slightly more practical level, by asking what would be different about a dynamical experiment performed in a room shielded from external light. Clearly, the light from the stars has no in-fluence on dynamics. One might postulate other influences which might be able to penetrate the walls of a room, influences having their origin in the matter of the stars. But Newton does not make such postulates; his laws are purely *local* in character. This may be a failing in Mach's view, but it cannot be remedied by global arguments of a kinematic character.

The question is sometimes raised as to whether the recent-ly discovered microwave background radiation might provide an absolute frame of reference in place of Mach's fixed stars. This radiation fills all of space in such a way that its intensity appears to be uniform in all directions when viewed with a radio telescope in just one special state of motion, which could be called absolute rest (see Section 12.6). A telescope in motion relative to this special one will register a non-uniform dis-tribution of intensity. The claimed advantage of this reference frame is that it is determined locally and in a very simple way. It also happens to provide a correct frame for the application of Newton's Laws to a high accuracy, higher than can be achieved by reference to the fixed stars. However, the sense in which it provides an absolute standard of rest is exactly the same as the sense in which this is provided by the stars. In no sense does it replace Newtonian absolute space, for it is still possible to perform dynamical experiments in a laboratory shielded from the influence of the radiation.

An analogy may be helpful here. We can think of the

motions of bodies as the movements of actors on a stage. The backcloth provides a reference field against which those movements are determined, since in a totally darkened environment it would not be possible to distinguish between uniform motion of the actor or the spectator. To this scene we may add further groups of actors or mechanical contrivances which may or may not be at rest relative to the backcloth, and we can use these to act as a reference frame. Or we can fill the stage with luminout vapours, which may or may not be flowing across the scene, which too can play the role of a reference frame. The existence of these further material determinants of motion, convenient as they may be, does not change the role of the backcloth of dynamically absolute space. To remove Newton's absolute space one has to remove the backcloth entirely, and with it the Newtonian plot which our moving actors follow.

Up to this point Mach has followed much the same line of development as Berkeley. At this stage in the argument - although not strictly in the sequence Mach actually follows - a new point is introduced. This can be seen clearly in Mach's treatment of the bucket experiment. What, he asks, could we say about the result of the experiment were we to perform it using a bucket 'several leagues thick'? We do not have a definite answer, but it might be that in this case the massive bucket would define the compass of inertia, so that the water surface would indeed exhibit maximum curvature at the stage of maximum relative rotation. Clearly, the suggestion is that if we treat the dynamical system as a whole, in particular if we take into account the whole Universe, then the mass of the Universe defines in some way a dynamical inertial frame as if, crudely speaking, it were like a gigantic gyroscope. This idea is reinforced when Mach attempts to give a quantitative expression to Newton's first law in terms of the centre of mass motion of the Universe.

This argument now appears to involve the distant stars, which contain the bulk of the mass of the Universe, on a dynamical level. The phenomenalist philosophy provides an easy justification for the involvement of the Universe in this way. For, as Mach rightly points out, 'we cannot know how [a particular body] K would act in the absence of [bodies] A, B, C...' (The Science of Mechanics, p.281). Any statement of the first law must therefore be 'nothing more or less than an abbreviated reference to *the entire universe*' (ibid. p.286).

This shows also the limitations of phenomenalism. For what is being implied is that despite the fact that Newton's Laws do not make any explicit reference to the rest of the Universe, it is a valid task for a scientific investigation to show that

54

implicitly they do. A more modern approach would be to accept
that two quite distinct hypotheses are possible: namely, that
either dynamics is a local theory with laws independent of the
structure of the Universe, or that the theory of dynamics should
be constructed globally in such a way as to exhibit an interaction between the matter in the Universe and a local dynamical
system. In the latter case this interaction must be one from
which a local system cannot be shielded without altering its
dynamical behaviour. Of course the hypotheses are then compared
on the basis of the range of validity of the resulting theories.
Newton took the first route; in general relativity Einstein
attempted to take the second. The degree to which he was successful, and a consideration of variations of this approach, are
the subjects of the later part of this book.

Despite this absence of any very deep formal analysis of
his idea, Mach goes on to hint at some pointers towards the
general theory of relativity. He remarks that

> ... it [is] possible that the law of inertia
> in its simple Newtonian form has only, for us
> human beings, a meaning which depends on
> space and time.   (ibid. p.295)

In as far as this slightly enigmatic statement is explained at
all, Mach appears to be referring to the manner in which he believes the local inertial frame to be determined by the stars.
This determination cannot be direct, he claims, since a chance
variation in the velocities of two distant bodies by impact could
not instantaneously alter the laboratory behaviour of a dynamical
system. In a similar way, he argues, the course of physical
processes would not come to a halt if a cessation of the rotation
of the Earth were to prevent us from measuring their progress in
terms of the solar day. Nevertheless, as Mach rightly continues,
one 'must feel the need of further insight - of knowledge of the
immediate connections, say, of the masses of the universe' (ibid.
p.296). In particular:

> There will hover before [us] as an ideal an
> insight into the principles of the whole matter,
> from which accelerated and inertial motions
> result in the *same* way.   (ibid.)

This is, albeit in imprecise terms, a clear presentiment
of the general principle of relativity. It is ironic that when
Einstein 'glanced into the depths' to acquire for us this new

perspective, the ageing Mach appears not even to have been able to understand that Einstein was concerned with this same problem.

Chapter 5: The Space-Time of Classical Dynamics

The *natural history* of scientific ideas is determined by
its own internal logic to a much greater extent than a super-
ficial glance at the historical record might indicate. The logic
becomes more evident once one translates the history into a technical language of the present. In dealing with Newtonian dynamics it is necessary to distinguish between Newton's claims
regarding the structure of his theory, and the structure of
space-time actually appropriate to the theory. The former were
presented and criticised in the preceding two chapters; the
latter will be constructed in the language of space-time geometry
in the two chapters which follow. The first of these deals with
classical dynamics in the absence of gravitational forces, and
the second with Newton's theory of gravity. We shall see here
the necessity of a geometric formulation, and we shall find in
the sequel that Einstein's general theory of relativity emerges
quite simply and naturally out of this formulation.

In the preceding chapters we have seen how the requirements of dynamical theories have imposed a structure on the
collection of events which constitute space-time. We saw, for
example, that certain topological or continuity requirements must
be met for the existence of particle paths, and that the imposition of a differential structure on the space-time manifold
enabled us to consider velocity vector fields. The existence of

an absolute time, and the validity of Euclidean geometry for absolutely simultaneous events, seem to have been taken as intuitively prior to a statement of the dynamical laws. On the other hand, the existence of absolute space, which is in fact incompatible with Newtonian dynamics, and its replacement by a more appropriate structure in agreement with the observed behaviour of dynamical systems, has, as we have seen, occasioned much discussion and confusion. It was, apparently, Lange (1885) who first gave the correct expression to the additional structure required for Newtonian dynamics in the absence of gravity. But it was not until the extension of these ideas to include gravity in the general theory of relativity, and their development in the work of Cartan (1923, 1924) that these views began to gain significant attention, and it is not until very recently that these concepts have become completely clear.

In this chapter, then, we are concerned with the structure imposed on space-time specifically by Newton's first law, rather than with the subsequent description of the evolution of dynamical systems within that space-time. This is to be contrasted with, for example, the analysis in Chapter 3 of Newtonian gravitational theory on the assumption of the existence of an absolute space. In the present chapter we shall treat the case in which gravity can be neglected; in Chapter 6 we shall see what modifications are required to incorporate gravitational forces.

§5.1   *Inertial frames*

Recall that the deficiency in Newton's postulate of an absolute space is that this cannot be discovered by dynamical experiments since, according to Newton's laws, observers moving with constant relative velocity find the same behaviour for dynamical systems. And the deficiency in the kinematic approach of Newton's major critics is that it does not allow a meaningful statement of the second law, since the accelerations in this law have to be referred to a class of unaccelerated motions. What is required is something intermediate to these extremes of rigidity and flexibility, namely, a description of the class of unaccelerated trajectories of the first law in terms of space-time structure.

The inertial motions of bodies according to the first law are supposed to be 'straight lines'. If we confine ourselves to a space picture of the motion of particles there is no problem here, since space is taken to exhibit the properties of Euclidean geometry according to which the notion of a straight line can be defined in one of several equivalent ways; for example, we may consider a straight line to be the shortest distance between two points. However, since we have no way of constructing an absolute space, the appropriate arena for dynamics is not space, but space-time. But in space-time we have not yet equipped ourselves with a way of measuring distances between points at different times. In fact, it is impossible to impose such a distance, or 'metric' structure on the space-time of classical dynamics. For suppose that according to one inertial observer two points P and Q have zero spatial separation, and time separation $t(Q) - t(P)$ (Fig. 5.1). Then we can always find an

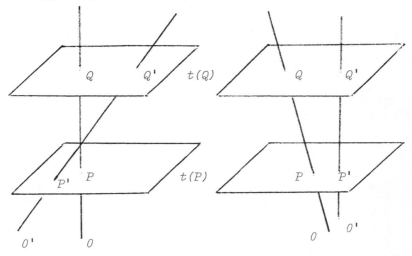

Fig. 5.1. By a permitted transformation (which consists of sliding the constant time slices over one another) from the inertial frame of O to that of O', the time separation $t(Q') - t(P')$ is seen to equal $t(Q) - t(P)$.

inertial observer such that any other pair of points P', Q' in $t = t(P)$, $t = t(Q)$, respectively, have precisely the same separation, hence precisely the same 'distance'. For we simply choose the inertial observer through P' and Q'. Thus all points with the same time separation are at the same 'distance', so at best 'distance' could have a trivial meaning of 'time separation'. However, we should then find that all absolutely simultaneous points have zero time separation, hence zero distance, which is clearly incompatible with the Euclidean structure of the three-dimensional spaces of constant time. Therefore we have no way of defining the distance between events which are not simultaneous,

so in Newtonian theory we certainly cannot define a straight
line in space-time as the shortest distance. Note here that we
are naturally confined to drawing pictures of space-time, which
does not have a metric, on pages of paper which do. Thus trajectories which appear to be straight lines on the printed page
may or may not represent our as yet undefined straight lines in
space-time, just as the printed maps of the London underground
railway represent as straight lines in the printer's metric,
paths which are far from straight for the engineer.

It follows from this discussion that we have to find a
way of describing mathematically the straight-line inertial trajectories postulated in the first law in a space-time which cannot carry a metric structure. To this end we can look at alternative definitions of straight lines in Euclidean geometry and
see if any of them can be suitably carried over to the present
situation. Thus we may consider (a) a straight line is a path
the tangent vector to which remains parallel to itself and (b)
a straight line is a path which in Cartesian coordinates, x^i,
($i = 1,2,3$), and absolute time, x^0, satisfies the equation

$$\frac{d^2 x^\mu}{dt^2} = 0$$

for some parameter t ($\mu = 0,1,2,3$). In (a) we add the concept
of parallel vectors to the space-time structure; in (b) we refer
to a special class of coordinate systems - something which we
should like to be able to do in a coordinate independent way!
It turns out that both of these generalisations are possible, and
indeed equivalent, and can be described in terms of an affine
connection (see Section 5.2). We turn first to a discussion of
parallelism.

Consider some region of space-time about a given point, P,
in which there is set up an arbitrary coordinate system of label-
lings of space-time points. The four vectors tangent to the coordinate axes at P constitute a possible *frame of reference* for
an (instantaneous) observer at P. For convenience we can take
three of the coordinates ($x^1(P)$, $x^2(P)$, $x^3(P)$) to label points
in the surface of constant absolute time through P and the fourth,
$x^0(P)$, to label surfaces of constant absolute time (in a continuous, strictly monotonic, but otherwise arbitrary manner). The
three tangent vectors (X_1, X_2, X_3) in the constant time surface
provide three reference directions in space, and the fourth vector, V, provides a reference velocity at P (Fig. 5.2).

In the three-dimensional surfaces t = constant, that is,
in ordinary space, we suppose that we know what it means to say

60

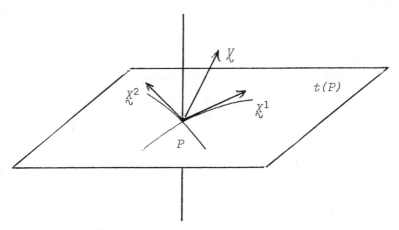

Fig. 5.2. A reference frame at P (one dimension, x^3, is supp-
ressed).

that a vector at some point P' is parallel to a given vector at
P. For the geometry of space is supposed to be mapped out by
ruler and compass according to the laws of Euclidean geometry.
Thus we assume that we can actually construct parallel vectors
in principle to any required accuracy, and hence, by constructing
at P' three vectors parallel to the directions specified by the
reference frame at P, that we can set up parallel spatial refer-
frames at absolutely simultaneous events. We have seen, however,
that the space-time of classical dynamics does not obey Euclidean
gometry since, in particular, the theorem of Pythagoras is not
satisfied. Therefore we have not yet given a meaning to the
parallelism of vectors representing reference velocities. We
need to look to the physical behaviour of clocks or the motions
of particles to do this.

Consider first a special case in which a trajectory through
P, tangent to the given reference velocity vector, V_P at P, of a
particle moving under no forces passes through the point Q, at
which it is required to define a vector parallel to V_P. We can
then define V_Q as parallel to V_P if V_Q is tangent to the given
trajectory at Q. In this way we use the physical inertial
motions to construct parallel vectors along a trajectory. In this
special case parallel vectors are those tangent to the same inertial path (Fig. 5.3).

More generally, consider a vector V_P at P not tangent to
the inertial path PQ. This represents the instantaneous velocity

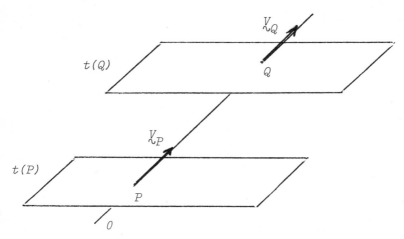

Fig. 5.3. $\mathcal{V}_Q$, tangent to the world-line of the inertial observer, O, at Q is defined to be parallel to $\mathcal{V}_P$, the tangent vector at P.

at P of a particle, as seen by the inertial observer moving from P to Q. We define the parallel vector $\mathcal{V}_Q$ at Q by requiring that it represents the instantaneous motion at Q of a particle with the same velocity as that at P according to the inertial observer. The parallelism of spatial vectors at P and Q could be defined as a limiting case of this construction, since a vector in the space section t = constant represents the path of a particle with infinite velocity. More directly, however, let $\mathcal{X}_P$ represent the distance and direction of a particle subject to no forces and at rest as seen by the inertial observer with world-line PQ. This particle will remain relatively at rest and its distance and direction at Q defines a vector $\mathcal{X}_Q$ parallel to $\mathcal{X}_P$. Finally, if P' is the point on the path of this particle simultaneous with P, then the velocity vector $\mathcal{V}_{P'}$, tangent to the path is defined to be parallel to the reference velocity $\mathcal{V}_P$ tangent to the inertial trajectory PQ (Fig. 5.4).

In this way, given a reference frame at a point P we can construct a parallel frame at any other point P'. Physically one can imagine rigid frames, or templates for the three-dimensional coordinate axes, being carried through space-time by inertial, hence non-rotating, observers. This extends the idea of the Euclidean physicist carrying his rigid ruler (instantaneously) throughout space in order to mark out the grid on which the histories of dynamical systems are to be charted.

62

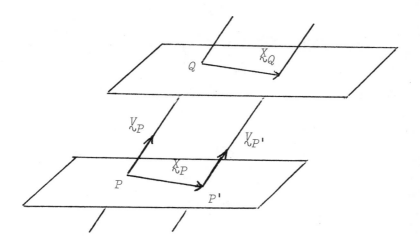

Fig. 5.4.   If inertial observers 0, $0'$ are relatively at rest
          then the vectors $\underset{\sim}{X}_P$ and $\underset{\sim}{X}_Q$ are defined to be parallel,
          as are the vectors $\underset{\sim}{V}_P$ and $\underset{\sim}{V}_{P'}$.

        Note that if the reference frame at P is constructed from
tangents to the coordinate lines through P, there is no reason
why the parallel frame at P' should consist of vectors tangent
to the coordinate axes, since we are free to change the coordinates at P' arbitrarily without affecting the physical situation
(Fig. 5.5).   Thus, in general, we want to consider fields of
frames on the space-time manifold that do not arise as tangent
vectors to a single coordinate system, even in a small region.
Of course, it may happen in certain cases that special coordinates can be chosen such that tangents to coordinate lines do give
rise to parallel frames, and indeed we shall find this to be the
case here.   But even at this stage it is helpful to allow for the
more general circumstances we shall meet when we deal later with
the presence of gravitational fields.
        There is now an important consistency requirement that
must be considered which is closely related to this point.   Let
PQ, $P'Q'$ be inertial trajectories in space-time of observers relatively at rest, and let P and P' and Q and Q' be pairs of absolutely simultaneous points.   Given a vector $\underset{\sim}{X}_P$ at P we can construct a vector at Q' parallel to $\underset{\sim}{X}_P$ in two ways: either we can
take $\underset{\sim}{X}_P$ to a parallel vector $\underset{\sim}{X}_Q$ at Q and thence construct a
vector $\underset{\sim}{X}'_{Q'}$ at Q' parallel to $\underset{\sim}{X}_Q$ at Q in the way we have des-

63

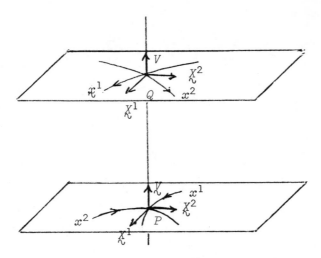

Fig. 5.5.   The frame at Q is parallel to that at P, but is not
tangent to the spatial coordinate axes at Q.

scribed;   or we can construct $\chi_{P'}$ at P' parallel to χ_P and take
this to a parallel vector $\chi_{Q'}$ at Q'. Both $\chi_{Q'}$ and $\chi_{Q}^{b'}$ are parallel to χ_P and must therefore be parallel to each other if parall-
elism is to be a transitive relation (Fig. 5.6).   This is our
consistency requirement and the guarantee that it be satisfied
is provided by Newton's First Law.   For consider the special case
where χ_P is tangent to the trajectory PQ.   Then $\chi_{P'}$ is tangent to
$P'Q'$ and since inertial observers relatively at rest initially
must remain at rest, χ_Q and $\chi_{Q'}$ are also tangent vectors.   But
then, by our construction $\chi_{Q}^{b'} = \chi_{Q'}$.   If more generally χ_P represents a velocity χ then it will follow from the construction
that both $\chi_{Q'}^{'}$ and $\chi_{Q'}$ represent the same relative velocity χ and
are therefore parallel as required.   This establishes that we can
talk about parallel vectors at separated points independently of
the path along which a vector is taken from one point to the
other.   For conciseness we speak of the *parallel transport* of a
vector along a path, meaning that we construct a vector at a
current point of the path parallel to a given vector at some
fiducial point.   We may say that in this case parallel transport
is independent of path.

This result is by no means a trivial one.   Suppose for
example we were to try to repeat the discussion taking instead

64

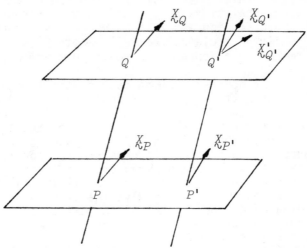

Fig. 5.6. If $\underset{\sim}{X}_{Q'}$ is parallel to $\underset{\sim}{X}_Q$ which is parallel to $\underset{\sim}{X}_P$ and $\underset{\sim}{X}'_{Q'}$ is parallel to $\underset{\sim}{X}_{P'}$ which is parallel to $\underset{\sim}{X}_P$ then for consistency (transitivity of parallelism) we must have $\underset{\sim}{X}_{Q'} = \underset{\sim}{X}'_{Q'}$.

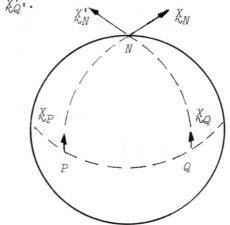

Fig. 5.7. On the surface of a sphere $\underset{\sim}{X}_P$ and $\underset{\sim}{X}_Q$ are parallel, as are the pairs $(\underset{\sim}{X}_P, \underset{\sim}{X}_N)$ and $(\underset{\sim}{X}_Q, \underset{\sim}{X}'_N)$, but $\underset{\sim}{X}'_N \neq \underset{\sim}{X}_N$.

of the space-time manifold the surface of a sphere equipped with the usual distance relations. By taking a vector round the triangular path shown in Figure 5.7 it is easy to see that here our consistency condition could not be met. Another example will be the space-time manifold appropriate to the Newtonian theory of gravity.

This consistency requirement is essentially equivalent to the possibility of setting up a coordinate system in which vectors tangent to the coordinate axes are parallel. Such a coordinate system is the non-metrical equivalent of the metrically defined Cartesian (orthogonal) coordinate system of Euclidean geometry. Suppose, then, that the consistency requirement is met, and hence that the inertial trajectories initially parallel remain so. We can then use a class of parallel trajectories to form the coordinate axes. Conversely, the existence of the special coordinates obviously implies the existence of path-independent parallel transport, since the parallelism defined by coordinate lines is path independent.

We can give a simple global picture of the structure we have discussed here if we introduce a new mathematical space called the *bundle of frames over space-time* or the *principal bundle*. Each point of this new space represents a point of space-time together with a frame of reference at that point (Fig. 5.8). Thus each point of the space is labelled by four

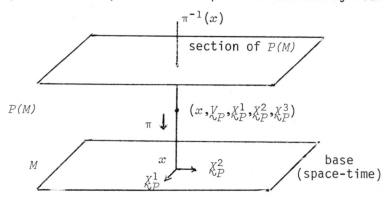

Fig. 5.8. The principle bundle $P(M)$ over M is a manifold each point of which represents a point of M together with a frame of reference at the point. $P(M)$ can be 'sliced up' into sections in a natural way (see text).

66

space-time coordinates together with the *4 x 4 = 16* numbers which describe the four vectors of the given reference frame in some space-time coordinate system. We can think of the space as consisting of a *base* manifold, which is just space-time, above each point of which are gathered all the possible frames of reference vectors at that point. The collection of reference frames at a space-time point x is called the *fibre* $\pi^{-1}(x)$ at x, and the space consists of a bundle of such fibres. It can be shown that the bundle of frames has a natural projection π on to the space-time manifold and a natural differentiable structure. Hence it is a differentiable manifold and this means that we can talk of moving smoothly from one frame to another. In particular we can discuss smooth curves in the bundle space and their tangent vectors.

Suppose now that we pick a particular frame at some point, x, in space-time M, hence that we pick a particular point in the principal bundle $P(M)$. At any other point in M we can construct a unique frame parallel to the given one. Hence we construct a unique point in any fibre associated with the given point. These points, one for each fibre, form a smooth *horizontal* slice of $P(M)$. We can repeat the procedure choosing a different frame at x from which to start. Thus we can think of $P(M)$ as sliced horizontally such that any point p in $P(M)$ belongs to just one slice. Technically, these slices are called *sections of P(M)* and the existence of global sections implies the existence of a horizontal projection. For any point $p \in P(M)$ can be projected horizontally on to a frame at some given (fixed but arbitrary) point $x \in M$; and, furthermore, this projection is *natural*, in the sense that it is independent of any arbitrary choice of coordinates, and hence physically meaningful. Since we already in the definition of $P(M)$ have the existence of a vertical projection down a fibre, it follows that $P(M)$ may here be represented as a product space, namely, as the product of space-time with the set of all reference frames at a point of space-time. We can look on this product structure as providing a natural choice of special coordinates, namely, the global Cartesian coordinates in which parallel vectors at different points have the same set of numerical values for their components.

To complete the discussion we can reverse the argument and construct the class of inertial observers of the First Law from a given horizontal sectioning of the bundle of frames. For, once a global section has been given, we can define the class of frames at points Q throughout space-time parallel to a given frame at some given point P; in fact, frames at P and Q are parallel if they lie in the same 'slice' or cross-section of the frame bundle.

Given a frame at P we can set off in a direction tangent to the reference velocity vector of the frame and arrive at a neighbouring point Q with a unique reference frame by parallel transport. From Q we make another infinitesimal step along the new reference velocity vector. In this way we construct, at least locally, an inertial trajectory through P and a non-rotating reference frame along that trajectory. For different initial choices of reference velocity at P, we construct locally all inertial paths through P. If we use the natural coordinates provided by the product structure then vectors tangent to corresponding coordinate lines will be parallel, since those tangent vectors have the same numerical components at all points. The coordinate mesh will in this case, therefore, provide a global inertial reference frame. This also serves to show that the above local construction can be extended to a global one.

Again one should note that this discussion is non-trivial: in the case of a sphere with its usual metric there are no continuous global sections of the frame bundle. For if there were then parallel transport would be independent of path, which we have seen is not the case. Thus for this manifold, at least, the construction is not possible.

We are now in a position to characterise the space-time structure of classical dynamics. We see that while, in contrast to Newton's views, space-time itself is not a product space, the bundle of frames of space-time *is* a product. This is the content of Newton's First Law.

§5.2   *The affine connection*

This structure can be described in an alternative local formulation as a special case of the concept of an *affine connection*, an idea which we shall need to discuss in greater generality later. The approach here connects also with the analytical formulation of the First Law as a differential equation, by allowing this equation to be written in an arbitrary system of coordinates.

Consider a coordinate patch in the neighbourhood of a point P. We may suppose a vector at P to be given in terms of its components with respect to the reference frame associated with the coordinate system at P. For any other point Q in this neighbourhood, we may ask, what are the components of the vector at Q parallel to that at P and expressed in terms of the coordinate reference frame at Q? The difference in these component values between P and Q will be zero in the simplest case that the reference frames are derived from Cartesian coordinates, since in these coordinates the components of parallel vectors are constant. In a general coordinate system, the components, V^α, will change by an amount, δV^α, which depends on the change in the coordinates between P and Q, δx, assumed small, and on the vector V^α itself. We can write

$$\delta V^\alpha \;=\; -\;\underset{\beta,\gamma}{\Sigma}\;\Gamma^\alpha_{\beta\gamma}\,V^\beta\,\delta x^\gamma \qquad\qquad (5.2.1)$$

where the $\Gamma^\alpha_{\beta\gamma}$ are the forty components of an object, Γ, called an *affine connection*, and the negative sign is an historical convention. It is conventional, and convenient, not to write the summation sign, Σ, explicitly, it being understood, unless otherwise stated, that sums are to be performed over all indices occurring twice in an expression. Thus, we write

$$\delta V^\alpha \;=\; -\;\Gamma^\alpha_{\beta\gamma}\,V^\beta\,\delta x^\gamma$$

for (5.2.1). This is not the most general expression one could conceive, but it correctly expresses the local version of our previous global construction. Specifying parallel vectors, therefore, is equivalent to specifying the components of an affine connection in some coordinates. The situation here is extremely special, since we know that by an appropriate choice of coordinates all the forty components of Γ can be made to vanish! We shall find that this is not the case when we come to include gravity in our dynamical system.

Parallelism of frames is, of course, described by repeating the above description four times for the four vectors constituting the reference frame. The formula may also be applied in the case that V^α is a velocity vector $\dfrac{dx^\alpha}{dt}$ that is required to remain parallel to itself at all times $t = x^O$. Then we have:

$$\delta\left(\frac{dx^\alpha}{dt}\right) \;=\; -\;\Gamma^\alpha_{\beta\gamma}\,\frac{dx^\beta}{dt}\cdot\delta x^\gamma \qquad ,$$

or

$$\frac{d^2x^\alpha}{dt^2} \;+\; \Gamma^\alpha_{\beta\gamma}\,\frac{dx^\beta}{dt}\,\frac{dx^\gamma}{dt} \;=\; 0 \qquad\qquad (5.2.2)$$

with summation over β and γ understood. Curves $x^\alpha(t)$ which satisfy this equation are called *geodesics*. In a frame (y^α) in which the components of Γ vanish, we have for the equation of

69

geodesics:

$$\frac{d^2 y^\alpha}{dt^2} = 0 \quad , \tag{5.2.3}$$

which is just the usual form for the equation of a straight line, and the trajectory of a body subject to no forces. Hence the geodesics here are inertial motions. The additional terms in (5.2.2) arise on using a non-inertial coordinate system and represent inertial forces. If we use a reference frame tangent to coordinates x^α then the expressions for Γ in these coordinates leads to expressions for the inertial forces in this non-inertial reference frame. If we know the relation between the x and y coordinates we can obtain explicit expressions for Γ. For, if $y^\alpha = Y^\alpha (x^\beta)$, then:

$$\frac{dy^\alpha}{dt} = \frac{\partial Y^\alpha}{\partial x^\beta} \frac{dx^\beta}{dt} \quad ,$$

and

$$\frac{d^2 y^\alpha}{dt^2} = 0 = \frac{\partial Y^\alpha}{\partial x^\beta} \frac{d^2 x^\beta}{dt^2} + \frac{\partial^2 Y^\alpha}{\partial x^\gamma \partial x^\beta} \frac{dx^\gamma}{dt} \frac{dx^\beta}{dt} \tag{5.2.4}$$

are the equations of geodesics. Using

$$\frac{\partial Y^\alpha}{\partial x^\beta} \frac{\partial x^\gamma}{\partial Y^\alpha} = 1 \text{ if } \beta = \gamma$$

$$\qquad\qquad = 0 \text{ if } \beta \neq \gamma \quad ,$$

we have from (5.2.2) and (5.2.3):

$$\Gamma^\alpha_{\beta\gamma} = \frac{\partial x^\alpha}{\partial Y^\delta} \frac{\partial^2 Y^\delta}{\partial x^\gamma \partial x^\beta} \quad . \tag{5.2.5}$$

For example, let the x coordinate system be derived from the inertial frame by constant acceleration g in the x^3 direction:

$$y^3(x) = x^3 - \tfrac{1}{2} gt^2 \quad .$$

Remembering that $x^0 = t$, it is easy to show that $\Gamma^3_{00} = -g$ and the remaining components vanish. Thus, the x^3 geodesic equation is

$$\frac{d^2 x^3}{dt^2} - g = 0 \quad . \tag{5.2.6}$$

70

Of course, by inspection, this is also the equation of motion of a body of mass m subject to a force mg, a statement of the obvious which will subsequently be of considerable significance.

In the space sections t = constant of space-time we have now apparently two definitions of a straight line. It is given in terms of a metric as the minimum distance between two points, and in terms of the affine structure as the solution of the equation of geodesics (5.2.2) or (5.2.3). Of course, we have anticipated that these definitions are equivalent. For Cartesian coordinates y^i (i = $1,2,3$) are defined by the statement that the square of the distance between neighbouring points y^i and y^i + dy^i is

$$ds^2 = (dy^1)^2 + (dy^2)^2 + (dy^3)^2 .$$

Minimising the integral

$$S_{PP'} = \int_P^{P'} ds ,$$

which gives the distance between events P and P' at the same absolute time, leads to parametric equations for the shortest path PP' of the form $\dfrac{d^2 y^i}{d\tau^2} = 0$ for some parameter τ. In the case t = constant, by following the argument leading to equation (5.2.3), we see that this is precisely the form of the equation for geodesics defined by the affine connection in the coordinates in which the components of the connection vanish. This is the required result, since we have shown that the two definitions agree in a particular coordinate system.

We might note that in a general coordinate system (x^i) we should have

$$ds^2 = \sum_{i,j,k} \frac{\partial y^k}{\partial x^i} \frac{\partial y^k}{\partial x^j} dx^i dx^j ,$$

where $y^k = y^k(x^i)$. Minimising the corresponding form of the integral for the distance along a path will yield the expression (5.2.5) for $\Gamma^i{}_{jk}$ on comparison with the appropriate three-space version of the geodesic equation (5.2.2). If we write

$$g_{ij} = \sum_k \frac{\partial y^k \partial y^k}{\partial x^i \partial x^j}$$

71

for the metric coefficients, the expression for Γ^i_{jk} can be written as

$$\Gamma^i_{jk} = \tfrac{1}{2} g^{il} \left[\frac{\partial}{\partial x^j} g_{kl} + \frac{\partial}{\partial x^k} g_{jl} - \frac{\partial}{\partial x^l} g_{jk} \right] ,$$

giving the components of the affine connection in terms of the metric. In cases where the space under discussion admits a metric, we shall see that this relation is generally valid.

The affine connection in space can therefore be said to arise from the Euclidean metric structure in the sense that both give rise to the same meaning for parallel vectors and geodesics. This is in complete contrast to the affine structure in space-time which exists in the absence of any possible metric structure.

Finally, it follows from the absence of a metric for the space-time of classical dynamics that this space-time cannot be the product of space and time as Newton had proposed. For space and time are individually metric spaces and the product of metric spaces is a metric space. A product space-time would therefore be a metric space, which we have seen to be false.

§5.3 *Galilean relativity*

The product structure of the frame bundle and, equivalently, the existence of the (special) affine connection correctly describe the structure of space-time corresponding to the presence of an observationally privileged class of inertial observers. The members of this class are equivalent with regard to the statement and verification of dynamical laws, and they are different from all other observers. This situation may usefully be described in terms of its symmetry, that is, in terms of transformations which could not be detected on the basis of the laws and structures postulated by the theory. Such transformations are called invariance transformations and physical systems are described as being - or not being - invariant under them. It will appear from the discussion (and Chapter 7) that, from the point of view of dynamical theory, the Lorentzian relativity of the special theory of relativity has a status equivalent to that of the Galilean relativity of Newtonian theory.

We have to distinguish between various types of transformations. For example, invariance under general coordinate transformations is required of any physically meaningful theory: the outcome of an experiment cannot depend upon whether the day on which it is performed is labelled according to the Gregorian or Jewish calendars. On the other hand it may depend on whether the experiment is performed at the Spring or Autumn equinox. This will give rise to a possible invariance which can be checked by observation or experiment. We distinguish also between a symmetry considered *passively*, which arises again from a change of labelling of the elements of a system, and a symmetry considered *actively*, in which a physical displacement of the system is envisaged. Finally, we shall consider the distinction between kinematical and dynamical symmetries. To avoid the more intractable philosophical issues that might arise, we shall for the most part restrict the discussion to the symmetries of dynamical theories, except for the purpose of illustrative examples.

In a sense it would be simplest if we could label points by proper names so that events and frames and velocity vectors etc., could keep their identity throughout any discussion. This would rid us of the need to keep track of the changing numerical values assigned to events in different coordinate systems. Unfortunately, this approach has not proved practicable, and in order to carry out computations we have to label points and objects with numbers depending on a choice of coordinate system or reference frame. Passive transformations are concerned with the freedom we have in changing this labelling.

Intuitively, the specification of a coordinate system consists in laying down a reference mesh over the region in question. We need a formalisation of this idea here for precision. Formally, since a coordinate system associates coordinates to points in a neighbourhood of some point P in a manifold M, we may think of it as a mapping ϕ from the neighbourhood U of P into a neighbourhood of a point O in R^n (if the manifold is n-dimensional - or R^4 if we restrict ourselves to the special case of four-dimensional space-time). We restrict ourselves to mapping small patches at a time since it is not possible in general to map the whole of M into R^n in a regular manner. (Consider for example a two-dimensional sphere for which stereographic projection on to R^2 breaks down at a pole.) This leads to technicalities concerning the smoothness of the overlap between different patches which need not concern us here. A transformation $r : R^4 \to R^4$ of coordinates in R^4 then generates a new coordinate system $\psi = r \circ \phi$ in a neighbourhood of P (Fig. 5.9).

Since the coordinate system can be chosen quite arbitrarily a physically meaningful result must be independent of this choice. Of course, one may make a special choice of coordinates for convenience of computation or discussion, but it must be clear that any result enunciated with this choice is in fact independent of

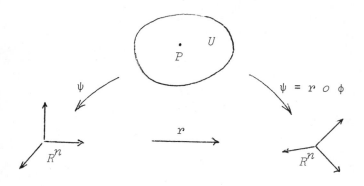

Fig. 5.9. The setting up of coordinates in a neighbourhood U of P may be thought of as a mapping ϕ of U to R^n. A transformation r of R^n generates a new coordinate system in U.

any arbitrariness. This is usually most simply guaranteed by expressing results in terms of locally observable coordinate independent quantities, such as, for example, a red-shift versus apparent luminosity relation for distant galaxies. Alternatively one may choose coordinates defined unambiguously by the physical system itself, which then have a physical significance. Here one might choose, for example, inertial coordinates defined by reference to the fixed stars. A theory may often appear to require a special system of coordinates, as does Newtonian dynamics in its conventional form. Here we appear to need to establish inertial coordinates for the statement of the second Law which can then only subsequently be transformed to an arbitrary coordinate system. This is, however, always indicative of an underlying additional geometrical structure in the theory, namely that which defines and is defined by the special class of coordinates. Making this structure explicit enables the theory to be written in a coordinate independent, or *covariant*, form. This preservation of form, although not necessarily of numerical content, is what is meant by the general covariance of a theory, and, as we have just seen, it can be achieved for any theory.

For example, in any coordinate system, inertial motions are described by equation (5.2.2), or

$$\frac{d^2x^\alpha}{dt^2} + \Gamma^\alpha_{\beta\gamma} \frac{dx^\beta}{dt} \frac{dx^\gamma}{dt} = 0 \qquad .$$

The values of the components of Γ will change under transformation of coordinates but not the form of the equation, which is therefore generally covariant.

We call this *general* covariance to emphasise that we are discussing form invariance under arbitrary (sufficiently smooth) transformations. We can, however, restrict our considerations to a more special class of transformations under which the complete structure of the theory is invariant, and which therefore expresses a physical symmetry.

To illustrate the point let us consider a simple example. Take for the manifold under discussion the surface of the Earth. Each point here is physically distinguished from any other point if not intrinsically, like the summit of Mount Everest, then at least through its surroundings. This characterisation is independent of the names we attach to places, or, more mathematically, of the zeros of latitude and longitude (and of the units used). If the Spanish pirate ships measured their longitude from a baseline through Madrid, this did not prevent them finding the English merchants reckoning from Greenwich (or vice-versa). It would be highly cumbersome, if at all possible, to state the accidents of history and geography in a form independent of some particular choice of coordinates. By making that choice in a physically defined manner, we ensure that our statements have physical substance.

Imagine now instead that the Earth were a perfect sphere. We can express this sphericity in terms of the distance relations that must be satisfied between points on its surface, namely, the laws of spherical trigonometry. More particularly, we can choose special coordinates, spherical polar coordinates, or, equivalently, latitude λ and longitude ϕ, adapted to the symmetry of a sphere. As we shall see more fully later, the geometrical relations are then contained in a statement of the distance between neighbouring points with coordinates (λ, ϕ) and $(\lambda + \delta\lambda, \phi + \delta\phi)$, which is $R\sqrt{\{(\delta\lambda)^2 + \cos^2\lambda \, (\delta\phi)^2\}}$, where R is the radius of the spherical Earth. In a different set of coordinates arbitrarily related to these, cartesian coordinates on a stereographic projection for example, the expression for the distance could take on a different, more complicated form. The geometrical relations would be the same in other coordinates, of course, and we could tell that we were dealing with the geometry of a sphere - at least in principle - since in principle we could find the transformation back to latitude and longitude in which the distance would again take on the prescribed form. On the other hand, λ and ϕ are not completely specified by this consideration since we can choose an equator and a meridian in many ways on a sphere. There are therefore many choices of coordinates for which the distance takes the special form, but this class of choices is distinguished as a whole from all other choices. The distinguishing feature of this special class is that it relates to the symmetry of the sphere. From any one special system one obtains any other by a suitable rotation of the coordinate mesh about some axis. The symmetry of the system resides in the fact that by geometrical methods alone one cannot tell whether this rotation has been performed. All members of the class of special coordinate systems are on an equal footing.

This discussion can be summarised by saying that rotation about any axis is a passive symmetry of the sphere. We do not imagine here that the sphere is actually rotated in any way, only that the labelling of points is appropriately changed. We can take the contrary view and imagine that the sphere is physically rotated from one orientation to another, and again we cannot tell by measurement whether this has occurred, whether, that is, we are now standing on a new region of the spherical Earth. This is expressed by saying that we have here an active symmetry.

These considerations can now be developed in the context of classical dynamics. Under a general (passive) transformation of coordinates we obtain a description of inertial trajectories in terms of a non-vanishing affine connection. The numerical description of an inertial path is thereby changed. However, we can specialise to that class of transformations under which the components of the connection remain zero. These transformations change the numerical expression of a particular inertial trajectory - for example, a body at rest may now be assigned a constant non-zero velocity - but they do not change the class of inertial paths as a whole. The labels attached to these paths are simply permuted amongst themselves. Therefore, by dynamical experiments, one cannot tell whether this permutation has been performed. These special transformations represent a symmetry of dynamical systems.

In the frame bundle picture these transformations may be characterised as taking one set of coordinates adapted to the product structure into another such set. For example, rotations of the axes of Cartesian coordinates in the plane R^2 manifestly preserve the product structure $R^2 = R \times R$, whereas the transformation to polar coordinates does not, since these are not adapted to display the product form.

We have already met this symmetry; it is just the invariance of Newtonian dynamics under the Galilean transformation. For, if the special transformations take some coordinates in which $\Gamma = 0$ into other such coordinates, they must leave invariant the equation for inertial trajectories in the form $d^2y^\alpha/dt^2 = 0$. The Galilean transformation may be regarded also as an active symmetry of classical dynamics. In this guise it leads to a statement of a Principle of Relatively in the form of a 'law of impotence'. For one cannot by dynamical experiment within a given system distinguish whether that system is at rest or in motion with constant velocity. This follows because according to the Galilean symmetry the class of trajectories in the one case is indistinguishable as a whole from the class of trajectories in the other. For the moment we leave open the question of whether by means of other types of experiment one can distinguish the two cases (see Chapter 7).

Physically, an active symmetry of space-time implies that we can displace a physical system from one location to another, or from one state of motion to another, and find that the results of all experiments conducted within the system do not distinguish the two cases. It should be noted that types of universal symmetry exist other than those associated with space-time. For example, there is a symmetry observed between the proton and neutron, at least as far as their non-electromagnetic properties are concerned, which can be described mathematically (and passively) as a rotational invariance in iso-spin space, and physically (and actively) as an identity of the outcome of certain scattering experiments when protons are replaced by neutrons.

A certain care is required in the interpretation of active symmetry. In particular, it is necessary to specify what is meant by the physical system. Clearly passengers on a train can determine whether or not they are moving by observing if they arrive at their destination, and the two possibilities are not equivalent! In this case the allowed experiments are those conducted entirely within the train, and the physical system does not include the environment of the train. The principle of relativity does not forbid the observation of the relative motion of different bodies, but only of the motion of a system as a whole without reference to external bodies. As a second example consider the observation of the period of an oscillating pendulum. There is no doubt that the pendulum will swing with a different period if transported to the lower gravity environment on the Moon. In this example the dynamical system consists of the Earth and the pendulum, and we must imagine them both to be transported to a new region of space under an active symmetry. There will also be effects on the motion of the pendulum due to the Sun, so we ought to take that along too. In this way we should end up relocating the entire Universe. What, in fact, is required is that the laws of physics should remain unchanged, not the accidents of environmental conditions. The active translational symmetry, or homogeneity, of Newtonian space-time reflects the identity of the laws which govern the motion of the pendulum on both the Earth and the Moon, and not the motion itself.

This approach is possible, however, only if we assume that the distinction between the fundamental laws and their particular applications has already been made. The concept of an active symmetry therefore depends on this distinction. For example, according to Aristotle the space around the Earth exhibits an intrinsic radial symmetry in the sense that all bodies fall with some velocity, independent of geographical location, towards, or away from, the centre of the Earth. This is a symmetry of space-time and does not depend on the presence of the Earth. Indeed, the central position of the Earth in the Universe can be explained, since heavy materials should accumulate at the centre of space. On the grounds that we see similar symmetry associated with other bodies, for example the falling of the Earth towards the Sun, we reject Aristotle's space-time symmetry in favour of an underlying intrinsic homogeneity, disturbed only by the accidental distribution of matter.

The action of symmetry transformations on dynamical systems is conveniently pictured as a transformation of the set of all possible trajectories of the constituent bodies into itself, either passively through a change of observer, or, actively, through a motion of the system. From the existence of these symmetries we may infer the existence of certain conservation laws which restrict the possible trajectories of dynamical systems. As an example, consider the symmetry of systems under time translation. Clearly we have the freedom to do experiments later rather than now, with the expectation that procrastination will not alter the results. This expectation may not be entirely justified since the overall expansion of the Universe may lead to differences even in dynamical experiments. However, we shall simply make the usual assumption that the expansion of the Universe, in accordance with the preceding discussion, plays no part in the laws of Newtonian dynamics. If we take dynamics to be invariant under time translation this means that dynamical quantities such as the energy, H, of a system cannot contain any explicit reference to time, but must be expressed entirely in terms of the positions, x_i, and velocities, $\dot{x}_i$, of the constituent particles. Therefore,

$$\frac{dH}{dt} = \sum_i \frac{\partial H}{\partial x_i} \dot{x}_i + \frac{\partial H}{\partial \dot{x}_i} \ddot{x}_i \ .$$

It can be shown that the Newtonian equations of motion for the particles can be written in the form

$$m_i \dot{x}_i = \frac{\partial H}{\partial \dot{x}_i} \ , \quad m_i \ddot{x}_i = -\frac{\partial H}{\partial x_i} \ ;$$

indeed, if the energy is expressible as a sum of kinetic and potential energies in the usual form, $H = \sum_i \frac{1}{2} m_i \dot{x}_i^2 + V$, these equations can be verified by direct computation. Using the equations of motion we find $dH/dt = 0$, or H = constant.

For any dynamically possible motion - a restriction which follows from the fact that the dynamical equations have been used - the energy of the system is constant in time. Or, equivalently, given the energy of the system at one time, the kinematic symmetry of time translation implies a restriction on the possible future dynamical behaviour. Similar computations relate the conservation of linear momentum to invariance under translation in space, and the conservation of angular momentum to rotational invariance. Note that we do not imply here that the system must be unchanging in time, or homogeneous in space, or spherically symmetric! This is immediately clear if we recall that we are dealing with symmetry transformations which transform trajectories into trajectories. For example, the system can have any shape at all, provided the privileged directions and positions so defined do not enter the dynamics explicitly to distinguish particular trajectories.

Conversely, however, not all systems appear to conserve momentum and energy. In these cases the dynamical evolution does not respect the kinematical symmetries of space-time. In accordance with our previous discussion this must arise from the concentration of attention on a subsystem of the total dynamical system. For example, the loss of energy due to frictional forces is not lost at all, but goes into kinetic energy of random motion in that part of the total system which provides the frictional force.

This particular example raises the interesting question of the *arrow of time*, detailed treatment of which is outside the scope of the discussion. The main point is that space-time seems to have a unique arrow of time which allows us unambiguously to distinguish the future from the past. Again in accordance with what has been said we should like to be able to relate this to physical processes occurring in space-time, not to an *a priori* intrinsic asymmetry of space-time itself. However, the fundamental physical laws appear to be essentially time-symmetric and this has led to difficulties for which there is as yet no agreed solution (see e.g. Davies 1974).

In special cases, systems may exhibit symmetry over and above that defined by the space-time structure. Indeed, such cases are often the only ones amenable to analytical solutions. For the extra symmetry implies invariance under a larger symmetry group and hence further restrictions on the dynamical possibilities. For example, a particular field of force, such as the gravitational field of the Sun, might be spherically symmetric. Under a rotation the set of possible trajectories of a particle subject to this force will then not only be mapped into itself, but certain trajectories will remain completely unaffected. In the spherically symmetric gravitational field, circular orbits will be transformed into circular orbits. These are additional dynamical symmetries. Such symmetries are, however, at best only approximate, in clear contrast to the space-time symmetries we have been discussing. Since a physical system will in general be subject to fluctuations or external disturbances, no sphere of gas can be exactly spherical, no planetary system exactly isolated. In practice it is therefore always important to discover whether the results of computations for an exactly symmetric situation bear any approximate relation to the physically realistic case.

Galilean relativity gives us a way of expressing the non-existence of an absolute velocity, and hence absolute space, at the same time as it introduces an absolute standard for acceleration into the theory. Consider the representation of the Earth as a smooth ellipsoid of revolution. Clearly the symmetry means that we could not tell if the Earth were to be rotated through some angle, yet we are supposed to be able to tell from its ellipsoidal shape that it *is* rotating, hence accelerating absolutely. There is here a mismatch between the geometrical symmetry and the dynamical laws, which is conceivably the motivation for much of the criticism of absolute acceleration in Newtonian theory. It appears to arise here, apparently, as an intrinsic property of space-time which is, indeed, the affine structure we have introduced. This affine structure expresses the existence of absolute acceleration in precisely the same way that a preferred vector field in Aristotelian theory expresses the existence

there of an absolute radial motion. The question arises, therefore, as to whether this apparently absolute affine structure
could in fact be ascribed to a material cause. The failure of
the dynamical laws to be invariant under an acceleration
transformation might then be accounted for if, as we have
discussed, the local dynamical system were only a subsystem.
Mach's ideas, as developed by Einstein, can be expressed by
saying that the complete system is the Universe as a whole, and
that the affine structure arises as a physical effect of the rest
of the Universe external to the chosen subsystem. In Einstein's
theory the nature of this physical influence is the gravitational
effect of distant matter. In the gravitational theory to which
we turn in the next section, Newton's inverse square law becomes
an expression of the dependence of the affine structure of space-
time on the material content. Einstein had hoped that his theory
would entail the complete determination of the affine structure
by matter. The extent to which this can be achieved is a
principle theme of the later part of this book.

Chapter 6:

Classical Space-Time in the   Presence of Gravity

We have seen that in Aristotelian theory the local privileged direction in the Earth's environment is attributed to a
property of space-time. In Newtonian dynamics the existence of
the preferred class of inertial observers is described in terms
of a product structure for the bundle of frames on space-time.
The new affine connection 'fields', Γ, are introduced to describe
this absolute motion; they do not eliminate it. We know that
the local behaviour of bodies near the Earth is better regarded
as a contingent physical interaction between the body and the
Earth, and that intrinsically space-time does not have preferred
directions. Could it be that the privileged status of inertial
motions could be eliminated in a theory in which privileged
frames, if they existed at all, were dependent upon matter, not
absolutely a property of space-time? We know that the elimination
of the Aristotelian structure required the introduction of new
physical ideas, principally the idea of universal gravity. We
may expect, therefore, that a successful elimination of the
absolutely privileged inertial motions would involve the introduction into classical dynamics of new physical ideas. This new
physics will again turn out to be, in fact, the gravitational
interaction.

If one simply rejects the Newtonian absolute, as Newton's critics were wont to do, and tries to start again in the construction of a dynamical theory, it is difficult to know where to begin. A more profitable approach is to look in detail at the postulates of Newtonian dynamics and ask whether there are any physical circumstances in which the validity of these can be doubted. Thus we try to press the theory to deal with circumstances for which it was not designed, and see whether it is adequate, or where it fails.

The absolute motion of Newtonian dynamics will be physically meaningful if we can always explicitly identify the class of force-free motions. Indeed, we explicitly remarked in Chapter 3 that the first of Newton's laws does not follow from the second, since the former requires that there should in fact exist identifiable force-free motions. If the forces in question in a certain system are all electromagnetic in origin, we can see what happens in their absence by switching them off. For example, we can refrain from kicking the ball, or we can switch off a power supply. Or we can repeat an experiment using electrically uncharged particles, which therefore do not respond to the electromagnetic interaction. These will provide us with 'straight line' trajectories against which the deviations of charges particles may be measured. Similar remarks apply to two of the other known types of basic forces, namely the so-called 'strong interaction' and the 'weak interaction'; in both cases there exist particles which are neutral with regard to these forces. Electrons, for example, do not interact with anything via the strong interaction. The only other known basic force of nature, the gravitational force, is quite different. There are no 'bodies' - by which we mean here any physical entities - which are gravitationally neutral. The gravitational field cannot be switched off and no object can be shielded from its influence. Of course, this does not mean that we can never do a laboratory experiment or discuss a problem without making explicit allowance for gravity. Gravity is a very much weaker force than the other three basic forces; at a practical level we need not always take account of it explicitly, since in many instances its practical effect is comparatively negligible. But at a fundamental level the effect of gravity may be decisive.

Not only are bodies always affected by gravity, but, as we remarked in Section 3.5, all bodies are affected in the same way independently of their structure and composition - at least, this is true if the bodies are sufficiently small that their presence does not perceptibly perturb the ambient field and such that they do not respond to gradients in the field. This experimental fact is incorporated into the theory as a postulate, hence as exactly true, as the *universality of free-fall*. According to this postulate then,

> the space-time trajectory of a small body
> subject to gravitational forces only, is
> the same as that of any other small body
> projected from the same space-time point
> (or strictly speaking a neighbouring point)
> with the same initial velocity.
> (Will, 1974, 1979)

For reasons we shall see later this statement is also referred to as the *Principle of Equivalence*, or, more precisely, as the *weak form* of the Principle of Equivalence. With this form of the Principle, and in what follows, we find that 'space-time' is not a superfluous luxury even in Newtonian physics.

It is just the universality of free fall that prevents us from constructing the Newtonian inertial frames, at least by dynamical observations, when gravity is present. To see this consider the simple Newtonian description of the fall of the apple. We wish to say that the apple falls because gravity acts. To do this we need to know the situation that would arise if gravity did not act. That is, in technical terms, we need to refer the acceleration of the apple to an inertial frame. What is intended is that the inertial frame should be one fixed to the Earth (or to the Sun, or the galactic centre if falling apples are to be described with great accuracy) in which objects subject to no net force maintain their state of rest. But we have seen that we cannot simply shield the apple from the gravity of the Earth in order to establish an inertial motion. Can we meaningfully assert that we somehow know what would happen if we could so shield the gravitational force?

This question is equivalent to asking whether we can determine a local inertial frame by some particular set of observations, since otherwise the assertion is clearly meaningless. By 'local' we mean here that the reference frame need only be set up over a small region of space for the description of dynamical behaviour in that region. What is meant by 'small' is left deliberately vague; we can and shall make it precise where necessary. Indeed, to do so is part of the problem we are posing. This restriction to local observations arises because Newton's laws are local in character: we change nothing according to the

84

theory by building ourselves a little nest around the apple and carrying out our observations there. The restriction to dynamical observations will be lifted when we consider general relativity - which, in a sense, arises from just the need to relax this restriction, there being more to the Universe than particle dynamics. It is required here because we are confined to Newtonian theory which does not encompass other phenomena in a consistent way (such as electromagnetic theory, etc.).

From our closed laboratory we consider various contrivances that will enable us to ascertain if we are in an inertial frame. For example, we drop a feather and a brick (both air-resistance free) with zero initial velocity and observe that they strike the laboratory floor at the same time. Now since they did not maintain their state of rest, we conclude we are not in an inertial laboratory. From the fact that they fall with equal uniform acceleration g, we conclude further that we are in a frame accelerating upwards at rate g with respect to an inertial frame. In fact we can assert that it is possible to interpret any other experiment in the same way. For we have already seen that the dynamical equation of motion for a body in a frame with acceleration $-g$ is the same as the Newtonian equation for a body of mass m subject to a force mg (Section 5.2). Since the outcome of any dynamical experiment is completely determined by the dynamical equations of motion, our assertion is proved.

This is not, of course, the outcome that was desired. What we had hoped to show was that we were at rest in a gravitational field of strength g. The point is that it follows from the Principle of Equivalence that there is in fact no meaningful distinction to be made *locally* between the two cases. For the universality of free fall means that, at least *locally*, we can always interpret a path as being the result of acceleration induced by gravity acting individually on the set of bodies, or as a result of viewing the system as a whole from a frame which is accelerated relative to an inertial frame. A distinction between these two possibilities can be made, as we shall see in due course, if we do not restrict ourselves to local observations. For the universality of free fall applies only to bodies projected from neighbouring points; bodies released from sufficiently widely separated points and observed for a sufficiently long time may reveal differences in the strength of gravity in their vicinities by falling with measurably different accelerations. But with the restriction to local observations there is no distinction between accelerated and unaccelerated observers, once we allow that gravity may be present, since the 'unaccelerated' observer in a gravitational field is locally indistinguishable from an

85

accelerated observer in the absence of gravity. In a correct theory of gravity, therefore, we expect all observers to appear on the same footing; we do not expect a privileged class in the sense of classical dynamics.

We have emphasised many times, however, that our dynamical theory requires us to begin by identifying the special class of inertial trajectories which can be taken to be straight lines. How are we to describe the falling apple without this? In a sense to be made precise by the full theory, the answer lies in turning Newton's problem round. The problem is not so much why the apple should fall to the ground, but exactly how it is prevented from so doing by the non-gravitational forces exerted by the tree! In this sense, it becomes the task of the theory to predict the strength of the non-gravitational forces required to cause specified departures from free fall or to describe the departures produced by given forces. Of course, to specify such departures it is also necessary to be able to describe the freely falling motions, but, from this point of view, gravity as a force disappears from the theory to be replaced by gravity as the geometry of space-time. Furthermore, this reformulation is appropriate not only to general relativity but equally to the Newtonian theory. The parallel with the Aristotelian explanation at this level of description is striking. Nevertheless, we shall see that at a deeper level, in which the structure so described is seen to be due to the presence of gravitating matter, this new viewpoint exactly expresses the content of Newtonian gravity. That we should arrive at a theory essentially equivalent in its quantitative predictions to Newtonian gravity is in itself a little disappointing, but we shall find that it provides the appropriate starting point for the relativistic generalisation. Indeed, there is no real reason why Newtonian gravity should not be taught correctly from this point of view in the first place, thereby removing any aura of impenetrable mystery that might still pervade the general theory of relativity.

§6.2 *The affine connection*

To reconstruct the theory of gravity we begin by considering the equivalence principle from another viewpoint. Since all bodies fall with equal acceleration locally in a gravitational field, then by falling along with the bodies with that same acceleration an observer sees those bodies not to be accelerated at all. Therefore, in the presence of gravity, we can observe bodies subject to no forces by falling freely in that gravitational field. We cannot distinguish locally between force free fall under gravity and no gravity at all. In this sense the role of the inertial frames of classical dynamics is taken over by the freely falling observers who are privileged to experience the local disappearance of gravity. It is from this point of view that the freely falling apple becomes the natural privileged motion, the progress of which is impeded by non-gravitational forces. Since free-fall, by definition, means motion in the absence of non-gravitational forces, in the absence of gravity too, *a fortiori*, free-fall means motion under no forces. That is, without gravity, free-falls are precisely the inertial tra-jectories; so the generalisation can be stated naturally that free-falls are in all cases the privileged paths.

Since we have asserted that in a certain sense we expect no privileged observers in the theory, it can be a source of con-fusion to see them appearing here in another sense. The sense in which we do not expect privileged motions is that of a once for all distinction between accelerated and non-accelerated observers as determined by local dynamics. We have just seen that Newtonian dynamics provides no way of making this distinction locally. However, given a particular distribution of gravitating bodies, there is a distinction between those paths that cannot be followed without the aid of non-gravitational forces, such as rocket motors, and those which can. The latter acquire, thereby, some other sort of privilege. These paths, moreover, change with a changing dis-tribution of matter.

To formalise the development of the theory so far we need a way of specifying this class of free-falls. Our experience with classical dynamics suggests we do this by means of an affine connection. In a generalised sense this will turn out to be correct, but there is a particular and crucial difference between the connection of classical dynamics and the connection in the theory of gravity. In particular, we shall find that in the presence of gravity the bundle of frames on space-time cannot be represented naturally as a product manifold. Thus, the presence of gravity modifies the structure of space-time even in the (neo)-Newtonian theory of gravitation. In the standard formulation of the Newtonian theory the product structure is retained by impos-ing it externally in a way which has nothing to do with the local dynamics, but which happens to give numerically correct results. For terrestrial problems we achieve this by taking a section of privileged frames to be those at rest with respect to the Earth. For celestial considerations we employ the background of the fixed stars. Of course, in the absence of gravity the current theory reduces to that of the previous discussion. It is now possible, in fact, to characterise the situations under which gravity can be neglected. This will occur if there is a negligible difference between the freely falling frame and the relatively accelerated frame which it is proposed to take to be the Newtonian inertial frame. The relative acceleration must then be small compared to the accelerations produced by other non-gravitational forces in the problem.

Recall that the appropriate mathematical structure in the non-gravitational case is obtained by using inertial observers to transport frames in a unique (parallel) way from point to point, except when the points in question are simultaneous when there is no inertial trajectory between them having finite velocity. In this latter case we use the Euclidean structure of space to define a connection. The Euclidean structure is supposed to be determined by geometrical measurement and is unchanged by the presence of gravity. For non-simultaneous events we can try to generalise the construction by using freely falling frames in place of inertial frames. Here we meet a difficulty. For consider free falls meeting at Q from initial points P and P'. This is certainly possible since apples dropped in Cambridge and Berlin could in principle collide at the centre of the Earth. Take frames parallel at P and P' and transport them to Q along the free-fall trajectories. The frames are initially parallel to each other and remain parallel to themselves along the trajectories *by definition*. But they are not parallel to each other at Q because, in the space-time picture, the time-components of the frames are not parallel (Fig. 6.1)! Therefore the attempt to set up an everywhere parallel system of frames in this way fails. Even if we consider the crossing of trajectories to be an exceptional case (which it is not), we can set up a contradiction by considering frames parallel at P, P' transported along PQ, $P'Q'$ to remain parallel to themselves, which will fail to be parallel to each other at Q and Q' (Fig. 6.2). The physically privileged motions do not define a 'horizontal' projection in the frame bundle which is therefore not a product in a natural way. (As we mentioned above, it can be considered as a product in many ways, just as Newtonian space-time can be taken to be a product in many ways; but no one of these is privileged or observer independent, hence such a product structure would not arise out of the dynamical theory.)

The problem arises because in a general gravitational field we cannot make a coordinate transformation to a frame in which *all* the free falls appear as straight lines. In the absence of gravity such a transformation was possible and we could use the fact that straight lines initially parallel remain parallel to obtain the product structure. In general the free falls will wind

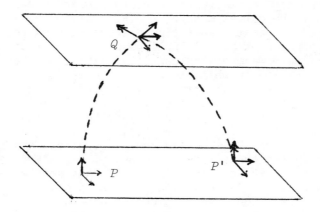

Fig. 6.1. Two frames at P and P' are each parallel to two
frames at Q but these two frames at Q are not parallel
to each other.

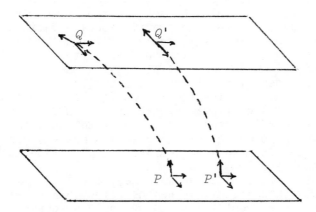

Fig. 6.2. Frames at Q and Q' are parallel to those at P and P'
respectively, but not to each other.

about each other and converge and diverge. We can slide the equal time slices of space-time so as to unwind the curves, but then if we further try to remove the convergence and divergence we shall find that the curves wind up again. This is made plausible by arguing that the unwinding operating fixes the three-spaces uniquely hence uses up all the freedom to make (active) transformations. Here we meet the non-local aspect of gravity.

There is one case in which the problem does not arise, namely that of a uniform gravitational field. According to the principle of equivalence, trajectories of free fall are here described by equation (5.2.6). By reversing the coordinate trans-formation leading to this equation we obtain (5.2.3), which rep-resents families of straight lines in the y-coordinate system. In this case, not only can the effects of gravity be removed locally, as demanded by the equivalence principle, but they can be removed over an extended region. The Equivalence Principle applies globally in this case. Conversely, since we have just noted that in general the effects of gravity cannot be removed exactly everywhere by choice of reference frame, we can see the need to emphasise that in general the equivalence principle applies only to small regions. Indeed, we can see that 'small' here should mean a region of space-time over which the gravitational field is sensibly constant within the accuracy of the measurements it is proposed to make. Over such regions the equivalence principle can correctly and profitably be applied to elucidate the situa-tion. Over larger regions experiments will reveal the existence of a true gravitational field through the convergence or diver-gence of free falls. These true gravitational effects are called 'tidal' effects since, in the Newtonian picture, they correspond to the forces which cause the tides on the Earth. The tides in fact arise from the difference in the gravitational pull of the Moon on the two sides of the Earth. These differences cannot be obtained in uniformly accelerated frames in the absence of gravity, and hence represent the effects of gravity which the theory must now explain. (Some differences in the gravitational field can be mimicked in non-uniformly accelerated frames but the general case cannot. This therefore adds a technical complication without altering the argument.)

The tidal effects and the way they reveal the existence of a gravitational field are simply demonstrated in Einstein's elevator though-experiment. We imagine again an observer enclosed in a featureless lift-cage, falling freely towards the centre of the Earth. He drops two apples which, of course, remain rela-tively at rest. The observer might conclude that he is an inertial observer in a region free of gravitational influences.

However, after a sufficient length of time, the tidal effect of the Earth's gravity will cause the apples to move measurably together. This is not caused by the mutual gravitational interaction of the apples themselves, which has so far been neglected, as a sceptical observer can verify by quantitative experiment. The observer will therefore correctly conclude that he is freely falling in a region affected by a centrally directed gravitational field. Thus one may again paraphrase this essentially Einsteinian view of the falling apple by saying that for one apple the departure from free-fall is ascribed to non-gravitational forces, but for two apples subject to no non-gravitational forces, the theory must now explain why they appear to approach each other in just such a way that they would meet at the centre of the Earth.

To describe this theory we seek something less rigid than everywhere parallel classes of reference frames. It remains true that between any two points at different times there is a unique free-fall, provided that the points are sufficiently close. For example, one could imagine a 'counter-Earth' freely falling in the opposite sense around the Sun which would meet the Earth twice in an orbit. (Recall that we do not need a space-time distance to discuss the relative separation of points.) Between these two intersections there are clearly two distinct free falls. For a pair of sufficiently close points we can determine a parallelism of frames by means of transport by freely falling observers. However parallelism defined in this way is not a transitive relation. By this we mean that if each of the vectors A and B is parallel to the vector C this does not imply that A and B are parallel. Intuitively this follows from the fact that initially parallel free falls - with tangent vectors, C, - may converge or diverge and give rise to non-parallel tangent vectors A and B.

Explicitly, we specify the parallel transport of a vector by giving, in some coordinate system, the change in the components of the vector $A^\mu(P) - A^\mu(Q)$ in transport from P to Q along a free fall trajectory as

$$A^\mu(P) - A^\mu(Q) = \Gamma^\mu_{\rho\sigma} A^\rho(P) \, \delta x^\sigma , \qquad (6.2.1)$$

where δx^σ is the coordinate difference between P and Q, assumed to be infinitesimal. There is now no coordinate system in which the Γ's may be made to vanish everywhere - for if there were the transitivity of the parallelism relation would be obvious from (6.2.1). Conversely, if parallelism were transitive, then the parallel transport of a vector round a closed curve would lead

91

back to the same vector as was initially transported. Any change in the vector is determined by the behaviour of Γ round the curve, and it can be shown that this change will be zero for every curve only if the Γ's can be made to vanish. This shows that the parallelism relation defined here is only transitive in the absence of gravity. It is sometimes convenient to use a co-ordinate system adapted to a given freely falling observer such that the coordinate frame is parallel along this trajectory. This is obviously possible since a freely falling observer considers himself to be at rest with gravity absent, and so rep-resents his own motion naturally as a straight line with an ass-ociated dynamically non-rotating frame. Formally, the coordinate system can be obtained by sliding the constant time surfaces over themselves until the particular trajectory is a straight line; it is clear that a rigid translation and rotation in each con-stant time slice is just the freedom required to achieve this, and is just the freedom available to us if we are to preserve the Euclidean structure of space.

We may also consider this structure from the viewpoint of the frame bundle. At a point in this bundle we are given a space-time event and a frame at that event. At any neighbouring space-time point we can find, by application of the affine re-lation (6.2.1) to each of the four frame vectors, a parallel frame, hence a particular neighbouring point in the bundle. In particular, displacements in space-time in four independent directions (three spatially orthogonal displacements and one in time) give rise to four displacements of a given frame in the bundle. Hence we construct four vector fields on the frame bundle which are nowhere tangent to the fibres $\pi^{-1}(x)$, $x \in R^4$, since they connect different space-time points. These are called horizontal vector fields (Fig. 6.3). The argument can be re-versed, so the specification of an affine connection is equiva-lent to the specification of these four vector fields. In the absence of gravity these vector fields certainly exist, and are (any) four independent tangent vectors to the surfaces in $P(M)$ which provide the privileged horizontal sectioning. With gravity present, the fields no longer join up to form global surfaces. Locally one may think of the vectors spanning an element of sur-face. But we know that different paths between the same pair of space-time points consisting, say, of segments of space-like and time-like geodesics will give rise to different pairs of parallel frames, hence paths between different points in the frame bundle. Consequently parallel transport of a frame round a closed curve in space-time does not lead to a closed curve in the frame bundle (Fig. 6.4). This means that the curve does not stay in a surface

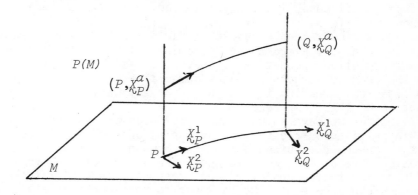

Fig. 6.3. Parallel displacement of the frame $\{\underset{\sim}{X}^a, a = 0,1,2,3\}$ from P to Q in M gives rise to vector fields on $P(M)$.

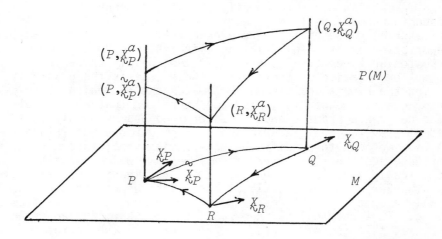

Fig. 6.4. Parallel transport round the closed geodesic triangle PQR takes $\underset{\sim}{X}_p$ to $\overset{\sim}{\underset{\sim}{X}}_p$ and does not in general give rise to a closed curve in $P(M)$.

in $P(M)$, hence that the elements of surface formed by the horizontal vectors at a point do not join together smoothly to form global sections. We see therefore that in the presence of gravity the frame bundle does not have a privileged horizontal sectioning defined by local physics, and so is not (naturally) a product manifold.

This result enables us to characterise the development of dynamics in a very elegant and economical manner: in passing from Aristotelian space-time to that of classical dynamics we lose the product structure of the space-time manifold; in passing from classical dynamics to Newtonian gravity we lose the product structure of the frame bundle.

So far we have shown that it is necessary to specify certain privileged free fall trajectories and we have investigated the consequences of so doing. We now know that gravitational motions are as they are because there are privileged vector fields on the frame bundle, which is essentially an Aristotelian or Copernican viewpoint. We have yet to consider the relation of this to Newton's universal gravity with regard to both the relation of free falls to Newtonian inertial motions, and the relation of the privileged fields to gravitating matter. We take up the former question in the remainder of this section and defer the latter problem to Section 6.3.

In fact in order to specify the affine connection, as opposed to asserting its existence, we have to give the Γ's in some materially identifiable reference frame. This will lead to a description of free falls in those frames. To do this we use the Newtonian kinematical inertial frames, namely those in uniform motion relative to the stars. Indeed, it is the possibility of using the kinematical frame that makes the Newtonian description of gravity meaningful, since we have seen that an unaccelerated reference frame cannot be determined unambiguously once we admit gravitational interactions. From our present (Einsteinian) view we have to specify the non-gravitational forces required to produce just that departure from free fall which constitutes kinematically inertial motion. That is, (at least approximately, since the Earth is not exactly at rest with respect to the fixed stars), we specify the force required to prevent the apple from falling. For spherical Earth of mass M, radius r, and an apple of mass m, this is, of course, GMm/r^2 away from the Earth. In general, $F = m\nabla\phi$ where ϕ is the gravitational potential satisfying Poisson's equation. Thus, relative to a freely falling frame the Newtonian inertial trajectory (the path of the apple at rest relative to the tree) is specified by

$$\frac{d^2 y^\alpha}{dt^2} = \frac{\partial \phi}{\partial y^\alpha} \quad .$$

Conversely, relative to the Newtonian inertial frame, the free fall under gravity must be given by

$$\frac{d^2 x^\alpha}{dt^2} + \frac{\partial \phi}{\partial x^\alpha} = 0 \quad . \tag{6.2.2}$$

Consequently, we can specify Γ by saying that there exists a coordinate frame in which the free falls take the form (6.2.2). For in an arbitrary coordinate frame we know that the equation for free falls takes the form

$$\frac{d^2 x^\alpha}{dt^2} + \Gamma^\alpha_{\beta\gamma} \frac{dx^\beta}{dt} \frac{dx^\gamma}{dt} = 0$$

(equation 5.2.2). Thus, in the coordinates in which (6.2.2) holds, we must have

$$\Gamma^\alpha_{oo} = \frac{\partial \phi}{\partial x^\alpha} \quad , \quad \text{(other components zero)}.$$

Therefore the assertion that there exists a frame in which Γ has this form serves to specify Γ.

There will be many such frames, namely all those in which the form of equation (6.2.2) remains unchanged, defined by the transformation

$$z^i = \sum_j R^i_j \, x^j + a^i(t) \qquad z^o = x^o \equiv ct \qquad \phi \rightarrow \phi + \sum_i \frac{d^2 a^i}{dt^2} x^i \tag{6.2.3}$$

where $a^i(t)$ are three arbitrary functions and R^i_j is a constant matrix representing a rigid rotation. This transformation corresponds to the freedom to make one (and only one) arbitrarily chosen trajectory a straight line, while preserving the Euclidean space metric. Note that this freedom is therefore greater than just a transformation between inertial frames. It is this that prevents one using this class of privileged frames to define a natural product structure; even though equation (6.2.2) does not change under (6.2.3), the form of Γ does. This is why, in this approach, the kinematical identification of a class of frames is at this stage still essential for the determination of

95

ϕ and hence Γ. In fact, as we shall see presently, (Section 6.4), the simple expedient of reinterpreting the equations of Newtonian theory in a logically satisfactory manner does not by itself lead to any fundamental change in the role of the kinematic inertial frame.

§6.3   *Tidal effects and curvature*

Before taking up the question of the relation of the affine structure to gravitating matter, we want to look at the way in which tidal effects enter the reformulated theory. The reason for this is that it is through the tidal effects that gravity is distinguished from accelerated motion, so one might expect that it is the aspect of the affine connection which produces tidal effects that will have to be related to the presence of matter. We shall see that here, just as in general relativity, the tidal effects of 'true' gravity are represented by a quantity called the 'curvature of space-time'.

The tidal effects are evaluated by comparing the behaviour of neighbouring free falls. If we adopt the point of view of an observer whose trajectory is one of those free falls, then this is represented by a straight line in his space-time picture, since this observer considers himself to be at rest and subject to no forces. The neighbouring free fall will converge or diverge according, physically, to the tidal gravitational force, and, mathematically, to the value of the affine connection. The amount of convergence or divergence, that is, the extent to which the neighbouring path is seen to deviate from a straight line, is determined by the degree to which the affine connection differs from zero, its value on the observer's trajectory in his coordinate frame. Thus, tidal effects are controlled by the derivatives of the affine connection.

For a precise calculation we use coordinates in which our observer's free fall has equation $x^i = 0$, and the coordinates of the neighbouring free fall are $(\delta x^i(t), t)$. These must satisfy the equation for a geodesic

$$0 = \frac{d^2 \delta x^i}{dt^2} + \Gamma^i_{oo} (\delta x^k) = \frac{d^2}{dt^2} \delta x^i + \delta x^k \left(\frac{\partial}{\partial x^k} \Gamma^i_{oo} \right)_{x=0}$$

(summed over k), where we have used Taylor's theorem to first order in the final term. If we write ξ^i for the geodesic separation vector parallel to δx^i (but of finite length), and write R^i_{oko} for $\partial_k \Gamma^i_{oo}$, then the equation becomes

96

$$0 = \ddot{\xi}^i + R^i_{oko} \xi^k \quad .$$

This is the "geodesic deviation equation" in Newtonian theory. It is unchanged in form under the coordinate transformations (6.2.3) which maintain the Euclidean metric. Of course, in a special frame where Γ takes the form $\Gamma^i_{oo} = \partial\phi/\partial x^i$, (other components zero), we have

$$R^i_{oko} = \partial^2\phi/\partial x_i \, \partial x^k \quad .$$

We write the quantity R^i_{oko} in this way since it is a special case of a quantity $R^\lambda{}_{\mu\nu\rho}$ which can be defined in a general frame of reference (and using a general time parameter), and which can be found by transforming the geodesic deviation equation to such a general coordinate frame. This quantity, R, is called the curvature of space-time by analogy with the corresponding quantity appearing in the general theory of relativity.

By looking at the preceding discussion in a more geometric manner we can see the way in which R represents curvature of space-time. Consider again the neighbouring free falls, but now focus attention on how a tangent vector V_P would change on parallel transport round $PQRSP$ (Fig. 6.5). V'_ξ is the differ-

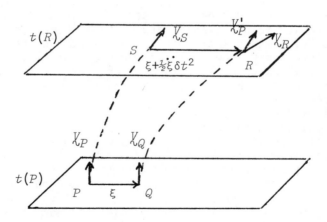

Fig. 6.5. *PS, QR* are free-falls with tangent vectors V_P, V_S, V_Q, V_R. V'_P is obtained from V_P by parallel transport along *PSR*.

ence between the velocities χ_P and χ_Q; take, for simplicity, $\dot{\xi} = 0$ initially, so χ_P and χ_Q are parallel. Since χ_Q transports into χ_R and χ_P into χ_S, $(\delta\dot{\xi})\delta t$ measures the change in initially parallel vectors, hence the difference $\chi_R - \chi_P'$ $(= \chi_R - \chi_S)$. According to the geodesic deviation equation this is given by $R^i_{oko}\,\delta t\,\delta\xi^k$ in the case $\chi_P = (1,0,0,0)$. In the special case depicted in Fig. 6.5, $\delta t\delta\xi^k$ is clearly the area enclosed by the curve $PQRSP$ to first order. In general, therefore, we expect the change in V_p to involve the components of an area $(\delta A)^{\nu\rho}$ (with two indices because it contains two directions). The change in V_p round an infinitesimal curve is given by

$$\Delta V_p^\lambda \equiv (\chi_R - \chi_P')^\lambda = -\tfrac{1}{2}R^\lambda_{\mu\nu\rho}\,V^\mu(\delta A)^{\nu\rho}$$

(The factor $\tfrac{1}{2}$ arises since R and δA are defined to satisfy $R^\lambda{}_{\mu\nu\rho} = -R^\lambda{}_{\mu\rho\nu}$ and $\delta A^{\nu\rho} = \delta A^{\rho\nu}$ so the summation on ν,ρ includes each term twice). The important point is that R measures the change in a vector on parallel transport round a closed curve. This change is non-zero if and only if R is non-zero.

Now, on an ordinary two dimensional flat surface the change in a vector on parallel transport round a closed curve is zero. This is because parallel straight lines, defined by means of the Euclidean distance between them, remain parallel, and parallelism is thus a transitive relation. On the other hand we can certainly find a closed curve on the surface of a sphere around which parallel transport changes the direction of a vector, as shown in Fig. 6.6. This example depends on the choice of great circles on the sphere as the analogues of free-falls - that is, as locally straight lines, and we shall have to wait until Chapter 8 to justify this as a 'natural' choice. However, it is certainly a possible choice, since these 'straight' lines can be obtained by projection from the plane. The point at present is merely that with this choice we have a clear analogy to justify the use of the appellation 'curved' to describe a space in which R is non-zero.

Thus, the space-time of the Newtonian theory of gravity is curved in just the sense described above of the behaviour of vectors under parallel transport. We can say that the affine geometry - the geometry that arises mathematically from the affine connection and physically from the behaviour of free-falls - is curved. On the other hand the geometry defined by the use of

98

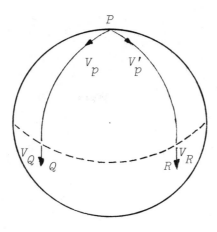

Fig. 6.6. Parallel transport round $PQRP$ takes V_p to $V_p' \neq V_p$.

measuring sticks is precisely that of Euclid and is quite separate from the affine geometry defined by dynamics. We shall
see that the great beauty and simplicity of general relativity
is that these two geometrical constructions become unified into
a single geometry of space-time.

The curvature R tells us everything about the behaviour
of neighbouring particles, hence about the effects of gravity.
To complete the story therefore one might try to express the
generation of tidal forces by gravitating matter in terms of
the generation of a non-zero R. Knowing R from the given or
observed distribution of matter we could calculate ϕ, and hence
the behaviour of bodies acted on by gravity. In this way the
free-falls would be defined in terms of the active gravitational
effect of matter. In standard Newtonian theory, ϕ is determined
by Poisson's equation. Thus we wish to recast Poisson's equation as an equation for R in a general coordinate system. If
we could do this without introducing any new structure into
space-time, then the resulting theory could be said to be generally covariant in the sense discussed above. Conversely, if it
were to turn out that this could not be done, then we should
have to investigate what additional structure we should have to
introduce to express the active gravity of matter. In fact, in
order merely to state the equations, no new structure will be
needed; this result is then a useful guide when we come to the
general theory of relativity. In a frame in which $\Gamma^i_{oo} = \partial\phi/\partial x^i$,
Poisson's equation can be written as $R^i_{oio} = -4\pi G/c^2 \rho$, since

99

$R^i_{oko} = \partial^2\phi/\partial x^i\partial x^k$. Since Newtonian theory contains an absolute time we are supposed to be able to pick out the time-components of R in a covariant way. In fact, the equation can be made manifestly covariant in form by expressing it in terms of the velocity V^μ of an observer as

$$R^i_{\ \mu i\nu} = -4\pi G/c^2\rho\ V_\mu\ V_\nu \quad ,$$

since this reduces to the previous form when $V^\mu = (1,0,0,0)$ (provided we use the symmetries of $R^\lambda_{\ \mu\nu\rho}$, in particular $R^\lambda_{\ \mu\nu\rho} = -R^\lambda_{\ \mu\rho\nu}$, which can be derived from the definition of the curvature).

Note that R has six non-zero components, but is apparently restricted by only one equation - we cannot simply add the condition that R be the second derivative of a scalar ϕ, since this would be non-covariant, depending as it does on a choice of coordinate system. In fact, we saw that R arises as the derivative of an affine connection, and this imposes certain restrictions on R. These - the so-called Bianchi identities - and the Poisson equation serve as equations for all the components of R. In this way a slight generalisation of Newtonian theory has arisen so that, strictly speaking, this theory is not exactly equivalent to Newtonian theory without some additional structure. Nevertheless, for problems of local dynamics, the two versions of the theory are equivalent and we have arrived at a covariant form of Newtonian gravity. However, the generalisation has allowed the construction of cosmological models which serve as analogues of certain models in general relativity (Ellis, 1971). We shall not be particularly concerned with this aspect, since the only models of physical relevance are those which can be obtained from the general relativistic models in the limit that 'Newtonian' theory is a satisfactory approximation - a circumstance that rarely prevails in cosmology.

Nevertheless, we cannot neglect cosmology here entirely. For Mach's criticisms involve consideration of all the matter in the Universe and it is of central relevance to our theme to discuss how dynamical theories fare with regard to those criticisms.

§6.4   *Absolute acceleration*

We have succeeded in reformulating Newtonian gravity in such a way that we do not require *a priori* privileged motions determined by local dynamics. Could it be, then, that in this formulation we have eliminated absolute motion and that only the relative motion of bodies is relevant? Could it be that there is now no essential difference between the Copernican and Ptolemaic pictures of the solar system, and that we are free to consider the Earth at rest and the stars in relative rotation? For a rotating Universe and a fixed Earth we must still find a flattening of the poles and the precession of the Foucault pendulum. These effects must be controlled by the components of the affine connection Γ, since these describe all the effects of gravitational and inertial forces, which are the only forces at work here. In a frame in which the Earth is non-rotating, the rotating stars must produce, through the solutions of Poisson's equation or its generalisation, just that Γ required to flatten the poles and cause the pendulum to precess. The problem is not that the Γ's necessarily fail to do this; the problem is that they may or may not give the correct behaviour depending on an arbitrary choice of boundary conditions.

To see this, consider the gravitational effect of a rotating body in Newtonian theory. It is exactly the same as that of a non-rotating one of the same shape and mass distribution, since the gravitational effect depends on the distribution of the density of matter. A rotating system of stars would appear to give rise to much the same gravitational potential as a non-rotating one; indeed, exactly the same if the rotation does not produce a flattening of the stellar distribution! How then could a rotating Universe produce the forces to flatten the poles? The affine connection certainly does change in passing from a description with a rotating Earth to a non-rotating Earth, just as it changed in passing to the linearly accelerated system as in equation (5.2.5). But this change has to be put in by hand in the sense that Poisson's equation does not distinguish between the two cases. The acceleration is represented by a homogeneous solution of that equation (that is, a solution of $R^i_{\;oio} = 0$), and it is a matter of choice whether this is included or not. Thus, it is a matter of choice whether the poles of the rotating Earth are flattened, or the poles of an Earth not rotating relative to the stars should be flattened. Since we know what the correct answer should be by observation, we make a choice of boundary conditions correspondingly. Effectively we single out the kinematically privileged reference frame attached to the fixed stars and demand boundary conditions ($\phi \rightarrow 0$ at ∞) such that the homogeneous solution does not appear in this frame. Grünbaum (1957) has argued that the boundary conditions in general relativity provide an absolute reference frame for accelerated motion. We see here that it is not just in relativity that this occurs.

101

The kinematically privileged reference frame is an additional structure to be added to space-time in order to elicit the correct answers from the equations of Newtonian theory. It amounts to postulating a globally determined privileged section of the frame bundle (since an essentially unique reference frame is determined at each space-time point)! It makes no difference that the privileged frames are related to something as material as all the matter in the Universe, because there is here no causal connection but merely an arbitrary assertion. With this addition, the picture of Copernicus prevails and the acceleration of the Earth in its orbit is absolute.

Chapter 7: Space-Time of Special Relativity

So far we have been concerned with the implications of
particle dynamics for space-time structure, in particular with
the Newtonian theory. But particle dynamics is not the whole
of classical physics. In particular we need to be able to deal
with the generation of electromagnetic effects by the motion of
particles carrying an electric charge, and not merely their
response to forces of electromagnetic origin. And we need to
deal with the dynamics of light. The theory which achieves
both of these is, of course, Maxwell's electrodynamics, which,
moreover, reveals light as merely an aspect of electromagnetic
phenomena, that effect by which charges in motion signal their
motion to other distant charges. What then is the space-time
structure required for electromagnetic theory, and what, conversely, does the theory tell us about space-time structure?
Until 1905 it was accepted that the success of Newtonian theory
could be interpreted as a justification of the intuitive assumptions that had been made. Einstein's success in 1905 with the
Special Theory of Relativity demonstrated that a consistent
interpretation of electromagnetic theory required a different
structure for space-time; one with which Newtonian particle
dynamics was only approximately compatible for systems with
small velocities. Moreover, this incompatibility could then be
rectified by a small change in Newton's second law, which al-

tered the predicted behaviour of bodies at speeds approaching
that of light.

Since 1905 countless experiments and developments have
been witness to the fruitfulness of Einstein's insight. The
availability of fast particles has led to tests of the theory
beyond the realm of small corrections, and indeed the function-
ing of the particle accelerators which produce those particles
is itself a rigorous test of the theory. Today, the ideas of
relativity are woven into our understanding of the fabric of
the Universe from chemistry to quasars.

§7.1   *Relative motion and light*

It was and is a well-known fact that electrodynamical
experiments do not reveal a state of absolute rest since their
outcome seems to depend only on relative velocities. Suppose,
for example, we drop a bar magnet through a coil. From the
point of view of the coil the changing magnetic field as the
magnet passes gives rise to an electric field. This in turn
exerts a force on the conducting electrons leading to an electric current in the circuit of which the coil forms part. From
the point of view of the magnet it is the coil that is moving,
hence the conduction electrons. The magnetic field is constant
in time and there is no electric field to act on those electrons.
Nevertheless, Maxwell's equations predict that there will be a
force on the electrons due to their motion in a magnetic field
of precisely the same magnitude as the force previously ascribed
to an electric field, and the conduction current will be precisely the same. The descriptions of what is happening are
different for the two different points of view, but the observed
outcome, the physical manifestation of an electric current of a
certain magnitude, is exactly the same. The experiment cannot
reveal which of the coil or magnet is *really* moving. Nor indeed
have countless other experiments been able to do so.

This discussion is usually carried through in terms of
the behaviour of light. The reason for this is that it becomes
possible to discuss an electromagnetic phenomenon, namely the
propagation of light, without the need to introduce the mathematical machinery of Maxwell's equations. For, provided we are
considering systems large compared to the wavelength of light,
it is possible to focus attention on the behaviour of light in
the regime of geometric optics, that is, to treat it in terms
of rays, with little regard to its wave nature. This should not
be allowed to obscure the fact that it is the whole theory of
electromagnetism that is really under discussion.

Let us then consider the behaviour of light rays in the space-time of classical dynamics. According to electrodynamics there is a reference frame in which certain electromagnetic waves travel with a speed c ($\sim$ 3 x 10^{10} cm sec^{-1}). This speed, constructed out of electrodynamic constants of the theory having *a priori* nothing to do with velocities or light waves, is a prediction of the theory. The agreement with the measured speed of light helps to secure the identification of light as electromagnetic waves. To an observer moving with speed v, light should appear to have a velocity $c \pm v$. If, in a particular frame we find light to travel with velocity c, then this frame may be singled out as an absolute state of rest and velocities relative to it as absolute velocities. This is a local dynamical determination of an absolute frame, not an arbitrary kinematical one, for the observations can be performed locally on the moving light beam. As far as the space-time structure is concerned, we are apparently no longer free to slide and rotate the constant time surfaces over themselves according to the Galilean transformation. For there is now a projection on to absolute space defined by the frame at absolute rest (Fig. 7.1).

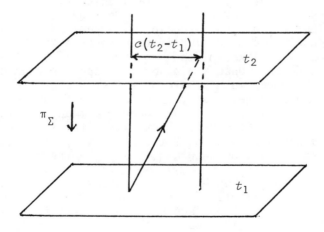

Fig. 7.1. Maxwell's electrodynamic description of light and Newtonian dynamics together fix the slicing of space-time and lead to a reinstatement of absolute space.

We have regained a product structure for space-time and with it Newtonian absolute space! Historically, this absolute frame could be regarded as the rest frame of the medium - the aether - through which the waves of light could be transmitted. The modern theory has no need for such a medium, but the argument is, however, totally independent of the existence or non-existence of the aether.

Clearly, the near identity of the predicted speed of light and the value we observe in the laboratory means that the Earth is moving slowly (compared to c) with respect to the absolute rest frame. This presumably explains the apparent relativity of electrodynamic phenomena, since for small velocities the absolute frame effects will be very small and could perhaps only be detected by sensitive experiments.

The Michelson-Morley experiment is only one, but perhaps the most famous, of such sensitive experiments. It was designed to detect the effect of the Earth's motion on the measured velocity of light. Since the velocity of the Earth changes in the course of a year by ~ 60 kms^{-1}, it cannot at all times be at rest relative to absolute space, so the measured velocity of light cannot always be the same. The great sensitivity of the experiment is achieved by not measuring this velocity directly but comparing it with the velocity of light perpendicular to the Earth's motion, which should remain unchanged. Note however that our account is a much simplified interpretation of the experiment. As is well known, the experiment failed to reveal the motion of the Earth. Once again absolute space eluded detection.

There has been much debate over the role of the Michelson-Morley experiment in relativity. There is certainly no direct reference to it in Einstein's 1905 paper, although he may have read of the result. Of course, Einstein, with his Machian education, was already convinced of the relativity of velocity, so there is no reason why he should give any prominence to just one example.

There has also been extensive discussion of the pedagogic utility of the experiment. The argument hinges on the possibility of construction of a perfectly rigid apparatus. The theory of relativity is often introduced with reference to perfect clocks and perfectly rigid rulers. However, this is no more than a device to allow the discussion to start. In the completed theory such objects are not required, the theory providing a full account of the outcome of any experiment. Indeed, since the construction of perfectly rigid bodies is not possible - in a rigid body the speed of sound would be infinite, which is impossible according to electrodynamics - it is contended that the

106

Michelson-Morley experiment is, in fact, merely a check on the construction of the equipment (e.g. Bondi 1967). This is in one sense completely true, and in another totally misses the point. If space-time had the Newtonian structure then a perfectly rigid Michelson-Morley interferometer could be constructed, and the absolute motion of the Earth thereby detected. Conversely the null result of the experiment is incompatible with the Newtonian absolute space-time structure, a result which manifests itself in many ways, including the non-existence of exact rigidity. In this sense, the experiment tests the foundations of relativity. On the other hand, once relativity is accepted as true, the Michelson-Morley experiment measures nothing other than the accuracy of construction of the apparatus, because it is part of the truth of relativity that the speed of light is constant, independent of the motion of the reference frame. And the Michelson-Morley experiment is not a confirmation of this truth only in the sense that according to relativity, the equipment must have been *built* so as to yield a null result. If the history of science had been different, and a compatible mechanics and electrodynamics had appeared together, the Michelson-Morley apparatus would indeed be no more than a (ludicrously oversensitive) rigidity meter. We have here a subtle example of how the interpretation of an observation depends on the theory!

We now face a serious problem: we have two theories, Newtonian mechanics and Maxwellian electrodynamics, which separately yield accurate predictions, but for which we have no appropriate space-time structure. It was Einstein's stroke of genius to see, as none of his contemporaries did, that what had to be done was to find a space-time structure appropriate to electrodynamics and then, if necessary, as was indeed the case, to modify Newtonian mechanics in such a way as to be compatible with this new space-time structure. There is no really good *a priori* reason for taking this approach; the fact of the matter is that Einstein alone tried it, and it worked.

Once again we have to begin by asking whether there is any structure in Newtonian space-time which goes beyond observational justification. The unobserved product structure arises from the assumed existence of a projection onto the time axis π_T, (or absolute time), together with a preferred speed for light. We have seen that dynamics provided no grounds for the assumption of a preferred projection π_Σ, or absolute space. Is there then any basis for the assumption of a privileged projection π_T? Or, equivalently, on what grounds do we ascribe absolute simultaneity to events? ·

Einstein gave a simple thought experiment which is des-

igned to elicit doubt about the reality of absolute simultaneity.
Imagine lightning to strike two ends of a moving train simultaneously as seen by an observer symmetrically positioned on the
railway embankment. To an observer in the middle of the train
the flashes do not appear to be simultaneous because the train
moves towards one, and away from the other, during the time in
which light is travelling from the event to the observer.

Of course, from the fact that he arrives in due course
at his destination, the train traveller can make the necessary
corrections to his observations so that he will agree with his
colleagues on the embankment. Or they can both make corrections
so as to agree with an observer at the galactic centre. Such
kinematically privileged frames can be chosen at will and have
nothing to do with the structure of space-time determined by
local dynamics. We can only assert a significant difference
between the traveller and the non-traveller if we can find a
local dynamical experiment which will give different results
for an observer in uniform motion and one at rest. Conversely,
if we wish to assert that there are no such differences - as
the Galilean invariance asserts for Newtonian physics - then it
would appear that we must give up the idea of absolute simultaneity. And, of course, the fact that experiments (such as the
Michelson-Morley experiment) designed to reveal the effects of
uniform motion fail to do so, suggests that we ought to assert
that there are none.

§7.2   *The Principle of Relativity and the Lorentz Transforma-*
        tions

The formal statement of the independence of physics including electrodynamics, from uniform motion is the Principle
of Relativity. This has already been stated for mechanical
phenomena. If Newton's laws are taken to be the laws of mechanics the Principle of Relativity leads to the Galilean transformation between inertial observers. Relativity itself was not
new with Einstein. The key development of the idea in the
Special Theory of Relativity is the extension of the Principle
from mechanical phenomena to all phenomena:

Principle of Relativity: it is impossible by local physical
                           experiment to distinguish a state of
                           rest or uniform motion.

The laws of physics must be the same for the class of inertial observers in uniform relative motion. The privileged

108

observers are unchanged from Newtonian theory. But we are no longer sure of the laws of mechanics, only those of electromagnetism! Therefore, we can find the transformation between inertial observers by requiring that these transformations leave Maxwell's equations unchanged. These are not the Galilean transformations, which was in fact the seminal observation that initiated Einstein's thinking. For by a Galilean transformation it is possible to reduce a light beam to rest leaving a spatially oscillatory electric field. This particular field, however, is not a possible configuration according to Maxwell's theory. Consequently the Galilean transformation and Maxwell's equations are incompatible, a conclusion which is simply a restatement of the above discussion.

Instead we are led to the Lorentz transformation either by consideration of the full Maxwell equations or, more usually, by arguing that a spherical wavefront of light must remain spherical when viewed from the relatively moving transformed reference frame. According to the Lorentz transformation, inertial observers are connected by the substitutions

$$x' = \frac{x-vt}{\sqrt{1-v^2/c^2}} \qquad t' = \frac{t-vx/c^2}{\sqrt{1-v^2/c^2}} \qquad y' = y, \quad z' = z$$

for motion of the primed frame with velocity v along the x-axis, and $t = 0 = t'$ taken conventionally as the time of coincidence of the origin of the two frames.

Note that at this point nothing new or mysterious has entered the discussion. The velocity has exactly the meaning it always has relating the same class of inertial observers, the distances are measured with rigid rods (in principle at least; the theory itself specifies how to make distance measurements in practice), and the time is that measured by standard atomic clocks. The Lorentz transformation correctly expresses the transformations we do, in fact, use in standard practice. For we do not correct our clocks to some standard absolute time depending on our absolute motion. It is not that we expect the velocity dependent correction to the rates of physical processes to be small; we expect them to be non-existent. We do not, for example, imagine the reactions in a rocket motor at $25,000$ ft sec^{-1} to be different from those in one on a test rig. We do not expect an intrinsic difference in a supernova explosion in a distant galaxy receding with 10% of the velocity of light. For none of us have ever met the master clock at absolute rest showing absolute time, and we do not feel ourselves obliged to correct our own clocks for their motion. The only mystery occurs in Relativity if one expects mechanical systems to behave as if Newton's laws were true at high relative velocities, for it has been shown many times by experiment that they are not. The apparent problems arise only if we assume that all (ideal) clocks in relative motion are equally good because they all keep absolute time - an idea which is compatible only with Newtonian dynamics - rather than because they keep their own time, as required by electrodynamics and relativistic mechanics.

It is easy to see that the Lorentz transformation implies a relativity of simultaneity. For events in the primed frame which are assigned the same time, $t' = 0$, do not occur at the same time in the unprimed frame. Any one inertial observer can divide spacetime into space and time thereby defining a projection π_T (and indeed a projection on to space π_Σ) by assigning the same time to events seen to be simultaneous (and the same position to points on the worldline of relatively stationary inertial observers). But these projections have no existence independent of the observer and are different for different observers. Indeed, there is nothing special about inertial observers here, for there are many other ways of imagining spacetime as a stacking together of arbitrary surfaces. Therefore, in passing from the space-time of classical dynamics to that of special relativity, we lose the privileged projection π_T and with it the idea of an absolute time. This is the sense of Minkowski's remark that

> Henceforth space by itself, and time by itself,
> are doomed to fade away into mere shadows, and
> only a kind of union of the two will preserve
> an independent reality.   (Minowski, 1908)

§7.3 *Minkowski's Space-time*

Minkowski's remark implies far more than the simple absence of the preferred projection π_T, for he was able to bring about a remarkable unification and simplification of the structure of space and time by viewing physics from the point of view of space-time.

Consider a clock moving along an inertial trajectory between points P and Q. From the point of view of the clock itself, it measures a time interval, $\Delta\tau$ say, between P and Q, and undergoes zero spatial displacement. This quantity $\Delta\tau$ represents a physical measurement independent of the coordinates, although its division into a space interval and a time interval does not. For,

110

from the Lorentz transformation formulae, for small intervals Δt, Δx, Δy, Δz, in the (t,x,y,z) coordinates appropriate to the time and space measurements of an observer having velocity v relative to the clock, we have

$$\Delta x \;=\; \gamma v \Delta \tau \qquad\qquad \Delta t \;=\; \gamma \Delta \tau$$

where, conventionally, $\gamma = (1 - v^2/c^2)^{-\frac{1}{2}}$. It is therefore obvious that

$$c^2 (\Delta \tau)^2 \;=\; c^2 (\Delta t)^2 - (\Delta x)^2 \quad,$$

since $v = (\Delta x / \Delta t)$.

It follows that the quantity $c^2 \Delta t^2 - \Delta x^2$ in this case represents a physical quantity, the time measured by a clock moving inertially between given points, and has a value independent of the coordinate system used to measure it, namely $(\Delta \tau)^2$. This quantity is called the *proper time interval* between the two points P and Q. It is obvious that in general, for relative motion not along the x-axis but in some fixed direction, the proper time interval is given by

$$c^2 (\Delta \tau)^2 = c^2 (\Delta t)^2 - (\Delta x)^2 - (\Delta y)^2 - (\Delta z)^2 \quad.$$

Suppose now we let our clock trace out a curve in space-time, $t = t(\lambda)$, $x = x(\lambda)$, $y = y(\lambda)$, $z = z(\lambda)$, where λ is some suitable parameter (which may be proper time τ). The tangent vector to this curve at the point P is $(dt/d\lambda, dx/d\lambda)$ at λ_P, (where we write $x(\lambda)$ for $(x(\lambda), y(\lambda), z(\lambda))$, in the (t,x) coordinates) or $(d\tau/d\lambda, 0,0,0)$ in the coordinates moving with the clock. Then

$$- c^2 \left[\frac{d\tau}{d\lambda}\right]^2 \;=\; - c^2 \left[\frac{dt}{d\lambda}\right]^2 + \left[\frac{dx}{d\lambda}\right]^2$$

evaluated at P, is a physical quantity which may be taken to be the negative of the square of the 'length' of the tangent vector at P. For all tangent vectors to inertial trajectories we can therefore associate a 'length'.

In the case when $c\Delta t < \Delta x$ for the separation of P and Q we should have $\Delta \tau$ imaginary, which clearly does not then represent the time measured by a physical clock. This is related to the causal structure which we discuss below. For such points we can find a frame (τ, ξ) such that $\Delta \tau = 0$, $\Delta \xi \neq 0$. $\Delta \xi$ is then called the *proper length* interval between P and Q, and is again

111

given by

$$+ (\Delta \underset{\sim}{\xi})^2 \; = \; - \; c^2 \Delta t^2 + \Delta \underset{\sim}{x}^2 \quad .$$

$\Delta \underset{\sim}{\xi}$ is just the result of applying Pythagoras' theorem in the space of constant time $\tau(P)$, (that is, the space-time separation obtained by using π_τ).   It is possible, but not trivial, to show how such an interval can be physically measured in the absence of exactly rigid rods (Marzke and Wheeler, 1964). We can use this to construct a length for vectors tangent to curves which are not possible inertial trajectories. This case justifies the use of the term 'length', since, in the appropriate rest frame, it is just the usual meaning of length.

In general then we are able to associate a length with a tangent vector, hence any vector, at a point. For a vector with components (V^0, V^1, V^2, V^3) the square of the length is just

$$\mid - \; c^2 (V^0)^2 + (V^1)^2 + (V^2)^2 + (V^3)^2 \mid \quad ,$$

where the bars denote the absolute value. *By definition*, this means we have a *metric* defined on the space-time. To relate to the Euclidean use of the term, as giving a distance between points along a curve rather than a magnitude for tangent vectors (the two are equivalent, of course), note that $\{\mid - \; c^2 (dt/d\lambda)^2 + (d\underset{\sim}{x}/d\lambda)^2 \mid\}^{\frac{1}{2}} \, d\lambda$ is just the distance between $P(\lambda)$, $Q(\lambda + d\lambda)$ and this expression can be integrated along any curve to give the associated 'length'.

The existence of a metric for space-time constitutes a fundamental distinction between relativistic and classical space-time. Any two points are separated by either a proper time or a proper length, but the computation of these quantities is not given by Pythagoras' theorem in general. In Euclidean geometry we deal with points and lines and planes and incidence relations together with a metric given by Pythagoras. In Minkowski space-time we deal with these same elements and relations, but with a metric given by the Minkowski expression $- \; c^2 t^2 + \underset{\sim}{x}^2$. Minkowski's geometry is our first example of a physically relevant metric geometry which is non-Euclidean.

§7.4   *Affine structure*

To proceed to describe a dynamical theory we need an affine connection. This will give us the motion of bodies under no forces, since these will follow straight lines as defined by

the connection. The easiest way to obtain a connection is to use
the one that is naturally provided by the metric. For example,
in plane Euclidean geometry the metric provides a specification
of distances and angles, and two lines remain parallel if they
are separated by a constant distance in a given direction. More
generally, the simplest way to visualise the relation between the
metric and the connection it provides is to note that the metric
takes a particular form (the Minkowski form, say) in a particular
class of coordinate systems, and the coordinate lines of these
systems provide tangent vectors which, by definition, remain
parallel along the lines.

It is not at all inevitable that the affine connection so
defined should be the one appropriate to a first law of relativistic mechanics. It may have happened that the metric appropriate
to the description of electromagnetic phenomena had nothing to do
with inertial motions - just as in classical dynamics the behaviour of measuring rods is quite distinct from the affine
geometry of inertial paths. However, experiment shows that it is
just those observers in inertial frames, that is, observers acted
on by no forces, who observe electromagnetic phenomena to obey
the laws of Maxwell. Relatively accelerated observers will, for
example, see light to be shifted in frequency. Now the construction above defines trajectories, the time-like coordinate lines,
of observers who measure the metric to take its Minkowski form,
hence who see Maxwell's equations satisfied. It follows that
these are also inertial observers for mechanical phenomena, and
we have defined the affine structure correctly.

This is a most important feature of relativity which can
be summarised by saying that the affine structure may be derived
from the metric structure. This is a situation common to relativity and Euclidean geometry, where at an elementary level no
attempt is made to distinguish the two, and is quite distinct
from the situation in classical dynamics. Indeed, this, together
with the fact that space-time now has a metric structure, may be
regarded as the key to the difference between Newtonian gravity
and general relativity; for we shall see that the arguments that
led to a reformulation of Newtonian gravity will lead straight to
general relativity.

§7.5 *The causal structure*

Points in the space-time that can be joined by light rays
satisfy the condition $(\Delta\tau)^2 = 0$, since in any Lorentz frame light
has speed c. At each point there is a cone of directions into
which light may travel. Inside this cone particles travel with

speeds less than c, outside with speeds greater than c. This describes a 'causal' structure for the space-time, dividing points into those which can and cannot be influenced by electromagnetic signals from a given event (Fig. 7.2). Were the speed

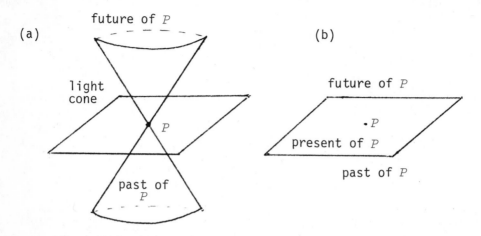

Fig. 7.2. The causal structures of Minkowski space-time (a) and Newtonian space-time (b).

of light to be infinite these cones would open out so that all simultaneous events in a given frame could be influenced by light signals, and simultaneity would become an observer independent concept. The causal structure of Newtonian theory is thus equivalent to absolute simultaneity and is appropriate to infinite signal velocities. In practice therefore Newtonian theory is approximately valid when a frame can be found in which all velocities, v, under consideration can be taken to be small compared to that of light. This frame is not unique, of course, but only errors of order $(c/v)^2$ are committed in taking one such frame to define absolute simultaneity. Note that the causal structure of classical dynamics is quite independent of its metric structure, whereas in relativity the causal structure is derived from the space-time metric.

The special theory of relativity certainly does not assume that nothing can travel faster than light, nor does it strictly predict that light travels at a limiting velocity. It does lead to a relativistic dynamics which is a simple generalisation of Newtonian mechanics, which shows that a body cannot be accelerated to superluminal speeds. And no-one has yet found a way of incorporating interactions into relativistic quantum mechanics which would consistently allow 'tachyons', as putative faster-than-light particles have been called, to interact with normal matter. Nevertheless 'particles' travelling on space-like trajectories are not necessarily self-contradictory. Since we could use such particles to travel back into the past, their behaviour must be constrained so as not to affect the future in an inconsistent way. We are familiar with the possibility that we may see a supersonic aircraft before we hear it and one might expect, if it proved necessary, to become familiar with the idea that a tachyon message may precede a visual signal. However light does play a special role in the theory of relativity in that the ordering of events separated by time-like intervals provided by light signals is invariant for all inertial observers connected by Lorentz transformations. This is true for no other signal velocity, and indeed part of the tachyon past for one observer may be part of the tachyon future for another. The special role of electromagnetic signals arises from the space-time structure required to support the theory of electromagnetism and relativistic mechanics - and indeed, at present, all fundamental physical theories - in just the same way as the space-time structure of classical dynamics was required to support Newtonian mechanics.

One might ask whether our apparent confidence in the validity of relativity could evaporate in the same way as belief in Newtonian theory did in 1905. Could special relativity be wrong? In this repsect it is obvious that relativity is no different from any other scientific theory. It will be proved wrong if the theory is pushed to new limits to which it turns out to be inadequate, that is, if new experimental regimes are revealed for which relativity gives incomplete or incorrect answers, just as Newtonian theory does when confronted with the electromagnetic phenomena. However, Newtonian absolute time *cannot* be reinstated by considerations of thought experiments for which the Maxwell's electromagnetic theory is postulated to be adequate. And furthermore no thought experiment can reveal any internal inconsistency in the theory of relativity just as none can be so discovered in Newtonian mechanics or indeed in Aristotelian dynamics. For at the level of internal consistency these theories reduce to pure mathematics, namely the geometries of space-time which have been the subject of our discussion. And these geometries are known to be internally consistent. Thus, we do not need to discuss either the twin 'paradox' in any of its guises or any other 'paradox', for these paradoxes are psychological only. Ask a well formulated question of the theory and it will give a definite answer. Any attempt to derive an internal inconsistency amounts to an attempt to prove the internal inconsistency of arithmetic.

§7.6   *Relativity and gravitation*

The theory of relativity becomes the 'special' theory only after it has been generalised. This generalisation consists of one thing, the inclusion of gravitational forces. This is achieved in the general theory of relativity to which we turn in the next chapter. It should be noted that while the special theory explicitly excludes gravitational effects it does not deal with uniform motion only, but with frames of reference in uniform relative motion from which bodies undergoing arbitrary accelerations can be viewed.

As a first attempt to include gravitation, one might ask how it is that the theory of electrostatics becomes part of relativity. The force between two charges e_1, e_2 at rest with separation r, is given by Cavendish's law $e_1 e_2 / r^2$, which has precisely the same form as the force between point masses in Newton's gravitational theory. How is it that one is a part of special relativity whereas the other involves an absolute time?

This paradox is resolved if we recall that the electric force is only one aspect of the electromagnetic field. Charges in motion experience in addition magnetic forces. We saw in our initial example of the magnet and the coil how the same force could be described as electric or magnetic in origin depending on the motion of the observer. Although the details are by no means simple, it is this effect that is the essential difference between electromagnetism and Newtonian gravity.

Could we achieve the relativisation of gravity by means of the addition of extra components to the gravitational force? We shall see that the gravitational force in general relativity does indeed have extra components depending on the motion of the interacting masses. It is possible, in some sense, to describe the theory from this point of view. However, this description is of dubious physical significance and there are crucial differences between electromagnetism and gravity, which prevent the analogy from being very fruitful in general. In particular, the gravitational field contains energy, equivalent to mass, which itself produces a further gravitational effect, whereas the electromagnetic field is not itself electrically charged, so does not produce further electromagnetic effects. Nevertheless it is sometimes possible and helpful to think of the difference between Newtonian gravity and general relativity in terms of extra magnetic type gravitational effects which appear in the latter theory.

Chapter 8: Space-Time of General Relativity

Special relativity arose because the space-time structure required by electromagnetic theory is incompatible with that of Newtonian dynamics. In a similar way, Einstein was led to general relativity because the space-time structure of special relativity is incompatible with that of Newtonian gravity. With hindsight the step from the special theory to the general theory can be taken with ease, since we have cast Newtonian gravity into a theory of space-time structure. We have simply to ask, and answer, the question: what difference does special relativity make to the geometric theory of gravity we have already found? The essential key to the enterprise is the so-called strong form of the Principle of Equivalence which is discussed in Section 8.2. This extends the Principle in its weak form (see Section 6.1) by requiring that *all* the laws of physics, and not just the laws of dynamics, should take the same form for an observer in free fall with or without the presence of gravity. Once this is accepted, and special relativity is regarded as the correct expression of local physics, then a theory of gravity based on a curved space-time geometry is inevitable.

Einstein, though, used more than just the equivalence principle to construct relativity. Of particular importance were the idea of covariance and the somewhat diffuse group of ideas he labelled 'Mach's Principle'. We have already discussed the covariance of physical laws as their invariance in form under changes in coordinate system (Section 5.3). Any physical law must be capable of covariant expression, and for this reason Einstein was later to regard covariance as being without physical content. However, Einstein's actual use of the idea of covariance was somewhat different from this. Essentially his use of the term corresponds to an observer-independence of physical laws. In this it relates both to the equivalence principle and to Mach's Principle. The relation with the equivalence principle arises because this provides the key to the equivalence of accelerated and non-accelerated observers. Such observers are distinguished from each other if one knows gravity to be absent; but by local dynamical experiments, according to the equivalence principle, one cannot arrive at a knowledge of the absence of gravity. Hence accelerated and unaccelerated observers appear locally to have the same status. This is, of course, precisely in accordance with the hopes expressed by Mach in the quotation from the Science of Mechanics (Section 4.4).

It is the use of the idea of covariance which led Einstein also to a theory of gravity based on a metric. For we saw that the affine structure of the space-time of classical mechanics is equivalent to the existence of privileged coordinate systems, namely those based on inertial motions. To avoid introducing such privileged coordinates the affine structure must not be introduced separately, but must arise from the structure already present. It was therefore natural for Einstein to take the connection to arise from the space-time metric.

Mach's Principle too plays a further role in the theory. We shall have much to say on this in the remainder of the book, but for the present let us take it that Mach's Principle states that inertial forces must arise from an interaction with distant matter. Now, according to the equivalence principle, inertial and gravitational forces are merely different aspects of the same thing, the distinction depending only on the state of motion of the observer. This enabled Einstein to identify the source of inertial forces as the gravitating matter in the Universe. It follows that the metric and hence the affine structure of space-time must be related to the matter content of that space-time. This is achieved through Einstein's field equations, discussed briefly in Section 8.5. They play a role in general relativity precisely analogous to that of the Poisson equation in Newtonian theory. The hope, however, was that the analogy would be incomplete, in that the extra components of the gravitational force in relativity, the magnetic-like components, would ensure the equivalence of, say, a rotating Earth in a fixed universe and a fixed

118

Earth in a rotating universe. For in the latter case the rotating stars might produce the extra forces needed to flatten the Earth's poles and cause the Foucault pendulum to precess. Since the form of the theory is supposed to be observer independent, one might wonder how this can fail to be true. In fact, the theory does fail here, and in exactly the same way that Newtonian theory does. Thus, the solution of Einstein's field equations, which exhibit the influence of the rotating universe, can only be obtained if suitable boundary conditions are postulated. And in the correct setting of those conditions, an absolutely unaccelerated reference frame again enters the theory.

Einstein was quick to realise that the theory fails also, in a more drastic way, to incorporate the initial motivating Machian ideas. Namely, situations are possible according to the theory - even if they do not actually occur - in which bodies exhibit their usual inertial resistance to impressed forces in the complete absence of any gravitating matter which might produce this inertial reaction. Thus the whole of the inertia of bodies is certainly not, in general, generated by the gravitational interaction of the stars, even if it might be in special cases.

In Chapter 9 these problems are discussed with reference to the special solutions of the field equations in which they arise.

§8.1   *The weak equivalence principle*

In special relativity, just as in classical dynamics, particles subject to no forces move on straight lines in space-time. In the special inertial coordinate systems such paths are described as solutions of the equations

$$\frac{d^2x^\mu}{ds^2} = 0 \quad ,$$

where now $x^\mu = (ct, \underset{\sim}{x})$ and s is the proper time along the path. More generally, in non-inertial coordinates y^μ, these same paths are represented by

$$\frac{d^2y^\mu}{ds^2} + \Gamma^\mu_{\nu\rho} \frac{dy^\nu}{ds} \frac{dy^\rho}{ds} = 0 \quad . \tag{8.1.1}$$

By carrying out the transformation from x to y coordinates, the components of $\Gamma^\lambda_{\mu\nu}$ are found to be

$$\Gamma^{\lambda}_{\mu\nu} = \frac{\partial y^{\lambda}}{\partial x^{\rho}} \frac{\partial^2 x^{\rho}}{\partial y^{\mu} \partial y^{\nu}} \quad ,$$

which is formally just the same as the non-relativistic expression. Thus in special relativity, as in classical dynamics, the motion of free particles is described by a special case of an affine connection - special in the sense that all its components can be reduced to zero everywhere by a suitable choice of global coordinate system.

In our introduction of gravity into the classical theory it was possible to observe at this point that the effects of Newtonian gravity could be transformed away, or introduced, at least locally, by passing to a description in terms of a non-inertial frame of reference. This proved to be both a weakness and a strength; on the debit side it showed that Newton's theory could not be considered satisfactory because inertial frames could not be defined locally in the presence of gravity; more positively, this equivalence of gravitational and inertial forces led us to a description of the classical theory of gravity in terms of an appropriate affine structure on space-time. In the present development, in which classical dynamics has been replaced by relativistic dynamics, we do not, of course, have a pre-existing theory of gravity to guide us to a space-time structure, since it is just such a theory that we wish to construct. It is therefore necessary to make some assumptions, suggested by the requirement that Newtonian theory should hold approximately for bodies travelling at much less than the speed of light, and ultimately, by experiment. Thus, we assume that the weak form of the principle of equivalence, the universality of free fall, will again be exactly satisfied in the new theory.

From the assumption of the universality of free fall we can proceed to show that gravity must be described by means of an affine connection, just as in the classical theory. For the assumption implies that by falling freely we eliminate the effects of gravity; we then use this class of free falls to define a notion of parallelism of reference frames and hence an affine connection. At first sight there appears to be a difficulty in that only a restricted set of space-time points can be joined by a free-fall, since bodies are restricted to move at less than the speed of light. In fact this presents no problems since arbitrary points can be joined by segments of free-falls even if some segments must be imagined to be traversed backwards in time. The extent to which this affine connection cannot be reduced to zero by choice of coordinates, and the notion of parallelism depends on path, corresponds to the extent to which the free-falls converge and diverge, that is, to the tidal effects of gravity.

At this level, then, the passage from Newtonian gravity to relativistic gravity is effected with scarcely more than a few changes of notation. Even the description in terms of the frame bundle would follow almost *mutatis mutandis* the account we gave in the classical theory. This is not, however, the complete story. For we have seen that in special relativity there is a remarkable unification of the metrical and dynamical attributes of space-time, in that the metric structure itself provides the affine connection appropriate to the description of inertial motions. Indeed the metric takes the special Lorentz form $ds^2 = -c^2dt^2 + dx^2 \equiv \eta_{\mu\nu}\, dx^\mu dx^\nu$, where $\eta_{\mu\nu} = \text{diag}\ (-1,1,1,1)$ and $x^0 = ct$, in precisely the inertial coordinate systems for which $\Gamma = 0$. In any other coordinate system (y^μ) we can find a direct relation between the components of Γ already given and the metric coefficients $g_{\mu\nu}$ defined by $ds^2 = g_{\mu\nu}dy^\mu dy^\nu$, and given explicitly by

$$g_{\mu\nu} = \frac{\partial x^\alpha}{\partial y^\mu}\frac{\partial x^\beta}{\partial y^\nu}\,\eta_{\alpha\beta} \quad .$$

It is an exercise in the manipulation of partial derivatives to verify that

$$\Gamma^\lambda_{\mu\nu} = \tfrac{1}{2}g^{\lambda\rho}\left(\frac{\partial g_{\rho\nu}}{\partial x^\mu} + \frac{\partial g_{\rho\mu}}{\partial x^\nu} - \frac{\partial g_{\mu\nu}}{\partial x^\rho}\right) \quad .$$

The correspondence can be expressed even more succinctly by noting that the free falls (8.1.1) are obtained as extremal curves, $y^\mu = y^\mu(s)$, of the integral

$$\int_{y=y(s)}\left|g_{\mu\nu}\frac{dy^\mu}{ds}\frac{dy^\nu}{ds}\right|^{\tfrac{1}{2}} ds \quad .$$

Since the integrand is just the length of the curve $y^\mu = y^\mu(s)$, this simply says that the free falls are the curves of longest proper time between two (causally connectable) events. The result that these are the longest rather than the shortest curves arises as a quirk of the Minkowski geometry; otherwise this is just the analogue of the result of Euclidean geometry that a straight line is the shortest distance between two points.

In Einstein's special theory the measurement of time is sufficient to enable us to predict the paths of free particles. We shall see that this same remarkable unification is transferred

121

also to the general theory.

The key to the unification of the metrical and affine
structures is a strengthened form of the Principle of Equivalence
which applies to all physical phenomena and not just the outcome
of dynamical experiments. According to this strong form of the
principle:

> All the non-gravitational laws of physics are
> the same in every local freely falling reference
> frame.

One might have thought that, even if local dynamical experiments
would not serve to define an inertial frame in the presence of
gravity, some other cunning experiments with, say, light, or
nuclear reactions, might be successful. The strong equivalence
principle asserts that this is not the case.
     In the stated form the principle does not commit us to an
opinion as to the true laws of physics. For our present purposes
we need to assume that the laws of special relativity represent
a valid description of physics in the absence of gravity. If, as
a matter of terminology, we call a frame in which special relativity is valid a *local Lorentz frame*, then the equivalence
principle states that the local Lorentz frames are freely falling
(and conversely). This is the version of the principle used by
Einstein and is often referred to as *Einstein's equivalence
principle*. It leads directly to a curved space-time theory of
gravity.
     One might object that special relativity has been tested
in laboratories on the Earth which are not in free falls. This
is indeed true, but experiments designed to test special relativity are arranged to take advantage of the weakness of Earth
gravity compared to other forces, and this means that departures
from the results of special relativity are made too small to be
of measurable consequence. On the other hand, one might try to
design equally cunning experiments to reveal precisely the effects
of a non-freely falling laboratory, and to find whether the frames
in which special relativity is exactly valid are those postulated
in the principle of equivalence. One such test is the gravitational red-shift experiment which we now describe (Will 1974).
     Suppose first (falsely, of course!) that the local Lorentz
frames have constant velocity with respect to the surface of the
Earth, and hence that we can set up a Lorentz frame which is att-

ached to the Earth despite the presence of the Earth's gravity. If this is so, then two observers, one, P, at the surface of the Earth, and the other, Q, at constant height h, are inertial observers since they are at rest in this frame. Let P emit light of frequency ν sec^{-1}. The time between wave-crests of this light is $\Delta t = \nu^{-1}$ sec. Refer now to the diagram (Fig. 8.1) on which are plotted the paths, AB, CD, of these crests between P and Q in space-time. These paths may not be straight lines, since they may be affected by the gravitational pull of the Earth. But there is no difference in the conditions of the experiment between AB and CD, so whatever the path AB between P and Q, whatever the effect of gravity on the propagation of light, path CD must be the same as AB. Therefore $AC = BD = \Delta t$. Since, by assumption, P and Q are at rest in a Lorentz frame, Δt is the proper time they both measure between the two wave crests. Consequently no frequency shift is observed.

The experiment has in fact been carried out and, not surprisingly, a frequency shift *is* observed. From this, one can deduce the acceleration, relative to the Earth, of the local Lorentz frames. Let us call this acceleration g. Then, relative to a Lorentz frame, P and Q traverse curved paths with acceleration g. From Fig. 8.2 we can see that BD is now greater than AC, and hence that Q observes light emitted by P to be red-shifted. To establish a quantitative relation requires an analysis of accelerated motion in special relativity to give the precise form of the paths P and Q. However, it is clear that the result arises essentially because in the time it takes light to travel from A to B the observer Q has been accelerated to some finite velocity relative to the emitter at A; consequently Q observes the light to be Doppler shifted. We can use this description to give an approximate Newtonian analysis of the problem which we might hope to give a correct order of magnitude estimate of the result. Thus for small accelerations the light travel time over a distance h from P to Q is $t_1 = h/c$ and the velocity of Q after this time at the moment of reception at B is $u_1 = gt_1 = gh/c$. Therefore the Doppler shift is $\delta\nu/\nu \simeq -u/c = -gh/c^2$ to first order in h, a result which is confirmed by a proper relativistic analysis. From the experimental results we obtain $g = 980$ cm sec^{-2} to an accuracy of 1% (Pound and Rebka, 1960). This is, of course, just the local acceleration due to gravity. So to an accuracy of 1%, and for this particular non-dynamical phenomenon, it is confirmed that the local Lorentz frames are freely falling as required by the strong equivalence principle.

Since we cannot verify the principle in this way for all possible experiments and all physical laws it would be satisfying

123

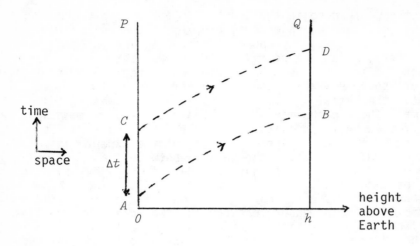

Fig. 8.1. Space-time diagram of two pulses of light between two observers P, and Q, at rest relative to the Earth.

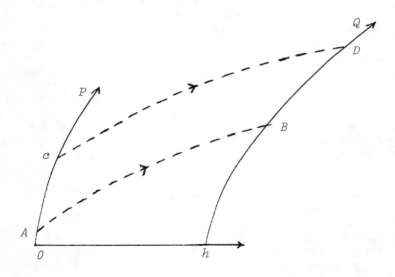

Fig. 8.2. The paths of the observers P and Q of Fig. 8.1. as seen from a Lorentz frame with relative acceleration g.

124

if it could be deduced theoretically from other postulates of the theory. The possibility that this might be so was suggested by Schiff, who conjectured that any theory which embodies the universality of free-fall necessarily satisfies the strong equivalence principle. Since we do not yet know *all* the laws of physics, no constructive proof of the conjecture can be given. However, it has been shown that for a certain class of theories, satisfying certain formal conditions, violation of the strong form of the principle would indeed entail violation of the weak form. The essential idea that makes this proof possible is that the dynamical behaviour of bodies is influenced by their internal structure: for example, the mass of a body will have a contribution from the energy associated with the forces that hold the body together. Now, for electromagnetic forces at least, it appears to be impossible to obtain a theory which violates the strong equivalence principle without at the same time causing an electrically bound body to violate the universality of free fall. One cannot, however, prove that if nuclear forces were to be included, which also violated the strong equivalence principle, there would not then be an exact cancellation of the contribution to violations of the weak form. On aesthetic grounds one might hope that such a possibility, if it exists, is not the theory chosen by Nature.

There is a further possible strengthening of the equivalence principle, sometimes referred to as the 'super-strong'form, and sometimes simply as the strong form, the name 'Einstein principle of equivalence' then being substituted for what we have called the 'strong' form. In this version the principle is taken to refer to all the laws of physics including those concerning gravity, instead of just the non-gravitational laws. This super-strong version is satisfied by the general theory of relativity, but not necessarily by all geometrical theories of gravity. The various forms of the principle of equivalence have been discussed by Will (1974, 1979), and experiments which can be regarded as indirect tests of these various versions have been discussion by Dicke (1964) and by Will.

§8.3 *Metric theories of gravity*

If we accept that the laws of special relativity are locally valid, in a freely falling frame, then a freely falling observer can assign lengths to vectors and hence to infinitesimal displacements in space-time. The freely falling observer can therefore associate a proper time interval between two finitely separated points on his world-line. In a neighbourhood of a point

P, which is sufficiently small that geodesics from P do not cross within the region, there is, by definition, a unique free fall connecting any point Q in the causal future or past of P to P. We can therefore assign a 'distance' $d(P,Q)$ to each pair of sufficiently close points P and Q by taking $d(P,Q)$ to be the proper time along the connecting geodesic. Since the geodesics are defined formally, in terms of the connection, even for points which are not causally separated we can also associate a distance, essentially the proper length, to spatially separated points. The set of numbers $d(P,Q)$ define the geometrical properties of the space-time. Since they have been obtained with the help of the connection, it is reasonable to expect some relation to hold between the geometry defined by the distances and the affine geometry of free-falls defined by the connection. To establish this relation we first find a more convenient way of expressing the content of the collection of distances $d(P,Q)$ in terms of metric coefficients.

If we take Q infinitesimally close to P we can replace the finite interval which depends on two points by functions defined entirely at a single point, P. It is more convenient to deal not with $d(P,Q)$ directly but with its square $\sigma(P,Q) \equiv (d(P,Q))^2$. If we let the coordinates of P be (x) and those of Q be $(x + \delta x)$, then

$$\sigma(x,x+\delta x) = \sigma(x,x) + \left[\frac{\partial\sigma(x,y)}{\partial y^\mu}\right]_{y=x} \delta x^\mu + \left[\frac{\partial^2\sigma(x,y)}{\partial y^\mu \partial y^\nu}\right]_{y=x} \frac{\delta x^\mu \delta x^\nu}{2} + \dots$$

To lowest order in the displacement δx the behaviour of the distance function must be just as it is in the Minkowski space-time of special relativity; this follows from the principle of equivalence. By direct computation it can then be shown that

$$\left[\frac{\partial\sigma(x,y)}{\partial y^\mu}\right]_{y=x} = 0 \quad,$$

since $\sigma(x,x+\delta x) \sim \eta_{\mu\nu}\delta x^\mu \delta x^\nu$ to lowest order. Obviously $\sigma(x,x) = 0$. We can write the remaining expression, correct to second order (and exact in the limit $\delta x \to 0$), as

$$- ds^2 \equiv c^2 d\tau^2 \equiv \sigma(x,x+dx) = g_{\mu\nu}dx^\mu dx^\nu \quad.$$

This is called the *line element*, or *metric*, and the $g_{\mu\nu}(x)$ are the metric coefficients. The expression is written as ds^2 to

bring to mind the square of an infinitesimal distance. From the
infinitesimal expression we can reconstruct the distance $d(P,Q)$
by integrating ds along a geodesic.

The development presented here can be achieved much more
economically with a little mathematical sophistication. For it
is, in fact, almost obvious that the assignment of infinitesimal
lengths by freely falling observers leads to a metric structure
for space-time of the above form. This follows because the observer who computes a distance, or proper time separation $ds^2 =
\eta_{\mu\nu}dy^\mu dy^\nu$ in a freely falling frame (y) will compute a distance

$$ds^2 = \eta_{\mu\nu} \frac{\partial y^\mu}{\partial x^\alpha} \frac{\partial y^\nu}{\partial x^\beta} dx^\alpha dx^\beta$$

in an arbitrary frame (x). At each point one may therefore pass
from the expression $ds^2 = g_{\mu\nu}dx^\mu dx^\nu$ for the metric to the Minkowski form by choice of frame at that point, although in general
this cannot be achieved throughout any finite region with a single
coordinate transformation.

We want now to establish the relation between the metric
coefficients and the connection. Both are known in a general coordinate frame (x), and in a freely falling frame, (y). A single
transformation from (x) to (y) must take us from the general expressions to the special forms, namely the Minkowski metric to
first order, and the connection with zero components to first
order. This can be achieved with the use of just one transformation, only if there is a relation between the objects transformed.
Consider then the general second order transformation

$$y^\mu = x^\mu + \tfrac{1}{2}\Lambda^\mu_{\nu\rho} x^\nu x^\rho \quad,$$

and its inverse

$$x^\mu = y^\mu - \tfrac{1}{2}\Lambda^\mu_{\nu\rho} y^\nu y^\rho + \ldots \quad,$$

in a neighbourhood of a point P at the origin of the coordinate
frames. The coordinates are restricted to be small quantities
here. The metric coefficients become

$$g_{\mu\nu}(y) = g_{\alpha\beta}(x) \frac{\partial x^\alpha}{\partial y^\mu} \frac{\partial x^\beta}{\partial y^\nu} \approx g_{\alpha\beta}(x) (\delta^\alpha_\mu - \Lambda^\alpha_{\mu\rho}y^\rho)(\delta^\beta_\nu - \Lambda^\beta_{\nu\sigma}y^\sigma)$$

or $\quad \eta_{\mu\nu} \approx g_{\mu\nu}(x) - \Lambda^\alpha_{\mu\rho}x^\rho g_{\alpha\nu} - \Lambda^\beta_{\nu\sigma}x^\sigma g_{\mu\beta}$

to first order in (x). Differentiate this expression with respect to x^λ, and let $(x) \to 0$ to obtain

$$0 = \frac{\partial g_{\mu\nu}}{\partial x^\lambda} - \Lambda^\alpha_{\mu\lambda} g_{\alpha\nu} - \Lambda^\alpha_{\nu\lambda} g_{\alpha\mu} \quad ,$$

which can be solved to give

$$\Lambda^\lambda_{\mu\nu} = \tfrac{1}{2} g^{\lambda\rho} \left[\frac{\partial g_{\rho\nu}}{\partial x^\mu} + \frac{\partial g_{\rho\mu}}{\partial x^\nu} - \frac{\partial g_{\mu\nu}}{\partial x^\rho} \right] \quad . \qquad (8.3.1)$$

On the other hand the geodesic equation in (x) coordinates is

$$\frac{d^2 x^\mu}{ds^2} + \Gamma^\mu_{\nu\rho} \frac{dx^\nu}{ds} \frac{dx^\rho}{ds} = 0 \qquad ; \qquad (8.3.2)$$

in (y) coordinates we have

$$\frac{d^2 y^\mu}{ds^2} - \Lambda^\mu_{\nu\rho} \frac{dy^\mu}{ds} \frac{dy^\rho}{ds} + \Gamma^\mu_{\nu\rho} \frac{dy^\nu}{ds} \frac{dy^\rho}{ds} = 0 \quad ,$$

on letting $(y) \to 0$. The correct form is

$$\frac{d^2 y}{ds^2} = 0 \quad ,$$

so we must have $\Gamma^\lambda_{\mu\nu} = \Lambda^\lambda_{\mu\nu}$. Thus (8.3.1) gives the relation between the components of the metric and the connection.

Having obtained the relation between the metric and the connection we can establish a very direct relation between the paths of freely falling particles and the measurement of proper time intervals. For it can be shown that a geodesic is the path which maximises the proper time between two fixed points. Thus, extremising the integral

$$I \equiv \int_P^Q \left| g_{\mu\nu} \frac{dx^\mu}{d\tau} \frac{dx^\nu}{d\tau} \right|^{\frac{1}{2}} d\tau$$

and using (8.3.1) gives precisely the equations (8.3.2) as the Euler-Lagrange equations for the extremal curve between P and Q.

As an example we can now explain why we stated in section 6.3 that the great circles on the sphere were the natural curves along which to consider parallel displacements. For with the usual metrical relations on a sphere the great circles are the geodesics, being curves which satisfy (8.3.2) when Γ is constructed according to (8.3.1). This follows without calculation since the great circles are the shortest distances between pairs of points on the sphere.

We have arrived at a metric theory of gravity: knowing the metric coefficients, or, equivalently, the proper time separation of events as measured by freely falling clocks, we can reconstruct the paths of bodies in free fall, that is, the paths of bodies subject only to gravitational forces. By a simple extension - the inclusion of a force term on the right hand side of the geodesic equation (8.3.2) - we can compute the paths of bodies subject to arbitrary forces. This is equivalent to saying that we have obtained a theory of dynamics in the presence of gravity. How we actually find the metric coefficients appropriate to any particular situation is a subject to which we shall turn when we discuss Einstein's field equations (Section 8.5).

§8.4   *Space-time curvature*

If we now go beyond the local regions in which the equivalence principle is valid we can see the true nature of the gravitational field revealed in the convergence or divergence of bodies in free fall. With some technical modifications this follows essentially as in the neo-Newtonian theory. For the vector connecting the two neighbouring geodesics in space-time, ξ^μ, we obtain the *geodesic deviation equation*, which has the form

$$\frac{d^2\xi^\mu}{ds^2} + \Gamma^\mu_{\lambda\nu}\frac{d\xi^\lambda}{ds}\frac{dx^\nu}{ds} + R^\mu{}_{\lambda\nu\rho}\,\xi^\nu\,\frac{dx^\lambda}{ds}\frac{dx^\rho}{ds} = 0 \quad . \qquad (8.4.1)$$

The parameter s is proper time along the fiducial geodesic, which has tangent vector dx^λ/ds. The quantity R appearing in (8.4.1) is related to the connection by

$$R^\lambda{}_{\mu\nu\rho} = -\frac{\partial\Gamma^\lambda_{\mu\nu}}{\partial x^\rho} + \frac{\partial\Gamma^\lambda_{\mu\rho}}{\partial x^\nu} - \Gamma^\lambda_{\sigma\rho}\,\Gamma^\sigma_{\mu\nu} + \Gamma^\lambda_{\sigma\nu}\,\Gamma^\sigma_{\mu\rho} \quad .$$

These relations hold whether or not the connection arises from a metric. If we do have a metric connection, as in general relativity, then R is called the Riemann curvature tensor of that metric.

The geodesic deviation equation (8.4.1) is a precise formulation of the notion that the deviation of freely falling bodies from parallel straight line trajectories expresses both the presence of a gravitational field and, equivalently, the curvature

129

of space-time. We noted in Section 6.3 the affine notion of
curvature as the change in a vector carried by parallel transport
round a closed curve. Precisely the same formal relations are
valid here. But now we have an additional notion of curvature
embodied in the metric coefficients. By this we mean no more
than that the relations between the proper times, $d(P,Q)$, between
various pairs of points P, Q, as measured by observers in free
fall, are not those appropriate to the Minkowski geometry of the
space-time of special relativity.

It is easy to give an analogy to describe the import of
this conclusion. On a flat plane one cannot construct four
points which are all equidistant from each other. On the surface
of a sphere this construction can be carried out in infinitely
many ways. The difference in the metrical relations expresses
the curved nature of the surface of the sphere.

In space-time such differences would appear in what one
might call a space-travel time-table of the Universe. For the
time taken to travel (at some constant velocity relative to distant parts of the Universe) from galaxy B to C will not in general
be related in a simple way to that required to travel from A to B
and from A to C. Indeed, in principle, the effect of the Earth's
gravity on the behaviour of clocks should appear in airline time-
tables, except that such time-tables would not normally be adhered
to with the requisite accuracy. Nevertheless, accurate experiments employing atomic clocks transported by air round the World
have shown the reality of such effects, and have confirmed the
quantitative predictions of general relativity (Hafele and
Keating, 1972). While experiments using galaxies are impracti-
cable for the foreseeable future, time-tables for the propagation
of light between some of the planets of the solar system have
been computed on the basis of general relativity and confirmed
by observation. Here one may think of light as consisting of
photons falling freely in the gravitational field of the Sun.

We might note that there is no sense in which one goes
outside of space-time to see its curvature revealed. The comparison with the surface of a sphere is misleading at this point, unless one bears in mind that the relevant geometrical relations on
the surface of a sphere are quite independent of its embedding in
a three-dimensional Euclidean space. To illustrate this point
further, observe that the surface of a torus acquires a metric
geometry from its embedding in three-dimensional Euclidean space
in which it appears as a curved surface. Nevertheless the torus
can be given a *different* intrinsic geometry which is locally the
same as that of the Euclidean plane. This follows because the
torus can be 'constructed' by glueing edges of a flat sheet of

paper. The construction cannot be carried out with a real sheet in three dimensions, but, although one cannot visualise it, it can be shown that the torus can be embedded in four dimensional space as a flat surface. (It is, in fact, the surface $x^1 = a\cos\theta$, $x^2 = a\sin\theta$, $x^3 = b\cos\phi$, $x^4 = b\sin\phi$ in four dimensional Euclidean space (x^1, x^2, x^3, x^4).)

§8.5 *The Einstein field equations*

The nature of a metric theory of gravity was expressed by Weyl in the terms 'space tells matter how to move'. This does not complete the theory since, to use the other half of Weyl's expression, we have not yet established that 'matter tells space how to curve'. In other words, we need an analogue of the Poisson equation which relates the Newtonian connection to the distribution of matter. At this point pure guesswork enters the development, constrained only by the fact that any equations we postulate must reduce to those of Newtonian gravity in those cases when Newton's theory is a good approximation. And, of course, the theory will only be acceptable if it can be made to agree with more refined observations, such as the bending of light by the Sun and the perihelion precession of the planet Mercury. There are, in fact, two different sets of equations, both proposed by Einstein, which can be described as the field equations of general relativity. The difference arises according as to whether a certain constant, the so-called cosmological constant, is put equal to zero. We shall take this up again in due course. Here we are merely making the point that there is no sense in which the equations governing the way matter curves space can be arrived at by deduction or proof. We can, however, provide plausible arguments for what might be the simplest adequate relations.

In Section 6.3 we saw that Poisson's equation could be written in covariant form as

$$R^i{}_{\mu i \nu} = -\tfrac{1}{2}\kappa\rho V_\mu V_\nu \qquad (8.5.1)$$

where $\kappa = 8\pi G/c^2$ and V^μ is the four-velocity of the gravitating matter at any point. We try to preserve this form as far as possible. However, the right hand side of the equation is unacceptable as it stands, since, according to special relativity, all forms of energy, not just rest-mass, contribute to the inertial mass of a body. From the principle of equivalence this implies that all forms of (non-gravitational) energy must contribute to passive gravitational mass, and hence, by the equality of action and reaction, to active mass. In addition, in special rel-

131

ativity, the flow of energy and momentum must be considered together. Let this be represented by the four-vector P^μ (as determined in a freely-falling frame). Now the flow of energy, and
therefore the value of P^μ, depends on the velocity of the observer; it is difficult to see how such a quantity could fit
on the right hand side of (8.5.1). However, there is, in special
relativity, another quantity, designated $T_{\mu\nu}$ and called the
stress-energy tensor, or the energy-momentum tensor, from which
the P^μ appropriate to an arbitrary observer with velocity V^μ may
be derived by

$$P^\mu = T^\mu{}_\nu V^\nu \quad .$$

Merely putting $T_{\mu\nu}$ on the right hand side of (8.5.1) does
not quite produce an acceptable equation. This is because there
is conservation of momentum and energy in special relativity,
which, in a local Lorentz coordinate frame is expressed by

$$\frac{\partial}{\partial y^\mu} T^{\mu\nu} = 0 \quad .$$

(This transforms to

$$\frac{\partial}{\partial x^\mu} T^{\mu\nu} + \Gamma^\nu{}_{\mu\lambda} T^{\mu\lambda} + \Gamma^\mu{}_{\mu\lambda} T^{\lambda\nu} = 0$$

in a general frame.) The left hand side of (8.5.1) does not satisfy this equation so cannot equal $T_{\mu\nu}$.

Let us write $R_{\mu\nu} = R^\lambda{}_{\mu\lambda\nu}$, where $R_{\mu\nu}$ is called the Ricci
tensor, and $R = R^\lambda{}_\lambda$ is the Ricci scalar. Then the expression
$(R_{\mu\nu} - \tfrac{1}{2} g_{\mu\nu} R)$ satisfies

$$\frac{\partial}{\partial y^\mu} (R_{\mu\nu} - \tfrac{1}{2} \eta_{\mu\nu} R) = 0$$

in a freely falling coordinate frame. Therefore, the equations

$$R_{\mu\nu} - \tfrac{1}{2} g_{\mu\nu} R = - \kappa T_{\mu\nu} \qquad (8.5.2)$$

are consistent with conservation of energy and momentum, and would
appear to be the simplest consistent modification of the neo-
Newtonian theory. It can be shown that they reduce to Poisson's
equations as a good approximation for weak gravitational fields,
motions at speeds much less than that of light, and media in which
the speed of sound is much less than the speed of light.

132

The equations (8.5.2) are Einstein's field equations. They are relations between the second derivatives of the metric coefficients and the distribution of energy and momentum. Since the metric determines the connection, and the connection determines the motion of particles, Einstein was led to hope that these equations might embody an expression of Mach's ideas, since they would appear to imply that the distribution of matter in the Universe determines the motion of bodies. From our point of view there is little difference between the situation here with regard to general relativity and the situation in Newtonian gravity, which we have already seen does not embody Einstein's interpretation of Mach's ideas. There is some difference, namely that Einstein's equations are more complicated in the sense that they allow more components of the gravitational interaction, which might give rise to inertial induction effects of distant matter. We shall see that, in fact, the equations in their stated differential form fail to achieve this objective. One might, however, bear in mind that even more complex equations, or possible alternative generalisations of the Poisson equation, could be more successful.

§8.6 *Consistency and completeness*

An important feature of Einstein's field equations is their non-linearity. By this we mean that the gravitational effect of two bodies acting together is not the sum of the effects that each would have on its own, since the presence of the one body alters the action of the other. This is quite different from the straightforward superposition of gravitational attraction in Newtonian theory, or the superposition of electrical attraction and repulsion in Maxwell's theory. And it has a more important consequence than merely making Einstein's equations difficult to solve. In electromagnetic theory the motion of a charged body is determined once the force on a charged particle is known. The nature of this force in its mathematical expression in terms of the electric and magnetic fields has to be posited quite separately from the equations which govern the generation of those electromagnetic fields. We have built up Einstein's theory on the claim that we know the motions of a body which merely responds to the ambient gravitational field without itself affecting that field; the body moves on a geodesic of the external field. Of course, such a situation cannot arise exactly in this way in practice, since all bodies have some gravitational influence. But we can idealise the situation by considering bodies so small that their effect is less than the preassigned accuracy of an

133.

experiment or observation under discussion. Such idealised
bodies are called 'test bodies'. Now the question obviously
arises as to how bodies which are not test bodies will respond
to the presence of other gravitating matter. Is it necessary to
postulate an expression for the force on the body in terms of
the variables which describe the total gravitational field,
namely the metric coefficients? The answer to this question
turns out to be negative and the contrast with electrodynamics
arises precisely because the equations of gravity are non-linear.

The detailed analysis of the motion of non-test bodies is
still a subject of active research (e.g. Ehlers, 1979). Schem-
atically, the procedure is to find a suitable sequence of successive approximations. At the nth stage of the sequence it is
assumed that the motion of the bodies is known to n^{th} order, and
that the next step is to compute the metric coefficients to
$(n + 1)^{th}$ order from the Einstein field equations. It turns out
that if these equations are to yield any solution at all, their
non-linearity implies that certain conditions must be satisfied.
These conditions are the equations of motion of the bodies to
order $(n + 1)$, and with these satisfied the Einstein equations
yield the metric coefficients to this order.

This result is important for the discussion of Mach's
ideas. For it means essentially that once we have determined
the metric from the matter distribution then we know how bodies
move. Thus the question of Mach's Principle becomes in general
relativity a question of the relation of the metric to the matter
distribution.

With respect to the motion of bodies the theory is complete and consistent. There is a further important aspect of
this completeness, namely that general relativity encompasses a
description of the whole of classical physics in the presence of
gravity, just as special relativity encompasses the whole of
physics in the absence of gravity. By this, we mean that the
laws of electromagnetic theory, say, are known in the presence
of space-time curvature, the behaviour of thermodynamic systems
can be predicted, and so on. The extent to which the theory is
complete with regard to the description of quantum phenomena is
not yet clear. Now it sometimes appears that the theory of general relativity is built up on the basis of certain assumptions
on the behaviour of clocks and light rays and perhaps also
measuring rods, which could be challenged and perhaps changed.
This is a false impression. For the completeness of the theory
has as a consequence that it completely *predicts* the behaviour
of clocks and light rays and measuring rods once the construction
of these is specified. The assumption upon which the theory

rests is the equivalence principle, in some form, and the only subsequent arbitrariness is a choice of the form of the field equations.

§8.7   *What is generalised in general relativity?*

The principle of special relativity states that all inertial frames are equivalent for the description of physical phenomena. We might therefore state a general principle of relativity in the form that all frames of reference are locally equivalent for the description of physical phenomena. This is valid, but there is an important sense in which it is not a generalisation of the special principle. In particular the class of coordinate transformations between general coordinate reference frames does not play a role analogous to the transformation between inertial reference frames in special relativity. For the general coordinate transformation does not imply an equivalence of the inertial and non-inertial reference frames of special relativity, but, effectively, a redefinition of an 'inertial' frame. The privileged frames in free fall in general relativity arise *a posteriori* from the curvature of space-time and do not represent a particular structure imposed on the space-time.

Nor is the general principle of relativity peculiar to general relativity; for Newtonian gravity, correctly presented, is compatible with the principle in much the same way that Newtonian dynamics is compatible with the principle of special relativity as it applies to dynamical (i.e. non-electromagnetic) phenomena. Indeed, we have seen how Newtonian theory is written without the introduction of special coordinate reference frames. It is therefore quite mistaken to think that the principle of special relativity singles out special relativity and that some generalised principle singles out general relativity. These theories are singled out by our faith in electromagnetic theory (and in the case of special relativity by a very large part of quantum physics). Thus the generalisation in general relativity consists just in the possibility of describing gravitation.

Chapter 9:

Solutions and Problems in General Relativity

A solution of the field equations of general relativity consists of more than just a knowledge of the metric coefficients in a local region. In principle one obtains also a global picture of the space-time. As a trivial example, suppose that in some context we are given a local expression for certain metric coefficients, not necessarily related to solutions of Einstein's equations, in the form

$$ds^2 \;=\; r^2(d\theta^2 + sin^2\theta d\phi^2) \quad . \qquad\qquad (9.1)$$

This is, locally, the metric of a sphere of radius of curvature r, with θ and ϕ the polar angles. The expression for the metric itself tells us this. If this expression is supposed to be valid in a neighbourhood of any point, then the manifold is also globally a sphere. Similar considerations apply to metrics obtained as solutions to the field equations. It is clear therefore that we cannot state the problem of solving the field equations in the form: from a given matter distribution find the corresponding geometry. For our example illustrates that before we know the geometry we may not be able to say whether the distribution of matter is even spatially finite, and we certainly cannot specify it in detail. Even at this level we see that the idea that the matter distribution determines the metric is not

136

without problems. This is essentially Weyl's (1924) objection to the suggestion that general relativity implements Mach's ideas.

A more immediately fatal objection arises from the fact that solutions of the field equations are possible which are generated by no matter at all. Einstein's attempt to modify the equations to exclude this possibility by the introduction of a so-called cosmological term was unsuccessful. Even if one regards the empty space-time geometries as inadmissible for some reason, there still exist solutions in which matter is present but for which the space-time geometry in certain regions, and hence the motion of freely falling bodies, is practically uninfluenced by the matter. For these solutions we have exactly the problem we found associated with the relation between a local dynamical and a kinematical inertial frame in Newtonian theory. In contrast to the case in Newtonian theory the behaviour of distant matter can *influence* the motion of particles, by the generation of effectively velocity or acceleration dependent forces, but the local particle trajectories cannot be completely determined in this way.

It is possible to take the view that the solutions we have been considering are appropriate only to local physics, and that the global view should encompass the whole Universe. In that case it can be shown that general relativity allows model universes with an arbitrary degree of rotation of the local freely falling frame relative to the motion of matter. In these universes centrifugal and Coriolis forces appear in frames fixed relative to the distant stars. Such universes may not be much like our own, but they are not ruled out by the theory of relativity (see Section 9.5).

Since the causal relations in space-time are determined by the metric, at places where the metric differs greatly from flat space-time, the space-time of special relativity, one may find that the causal relations too differ greatly from their intuitive form. Such objects as black holes may be considered in this light. There can arise causal pathologies involving horizons beyond which the motion of particles cannot be definitively predicted. In such cases the distribution of energy and momentum in one part of space-time would not determine the motion of bodies in another. Causal pathologies, such as closed time-like lines, may also be obtained, which would enable one to change one's past (Section 9.3). If such things are absurd it might be argued that a good theory should not admit them, and Einstein was much distressed by the discovery that the beauty of general relativity should be marred in this way.

If the motion of a freely falling observer comes to an end at a finite time we have the condition for a singularity, or edge, to the space-time. Here the curvature may increase without limit, but this is not necessarily the case. The existence of singularities, which is the generic situation in solutions of the field equations for realistic matter distributions, is an embarrassment in the theory, since it leaves the theory powerless to predict beyond them.

§9.1   *Mach's Principle and empty space solutions*

Despite the clear influence of Mach's ideas on Einstein's development of general relativity, as evidenced in Einstein's writings, we have seen that the theory can be constructed almost independently of these ideas. The only point at which they enter the discussion is in the requirement that the metric be related to the energy and momentum of matter alone through the field equations, and not to some other quantities in addition. Indeed, we can take Einstein's statement of what he referred to as 'Mach's Principle' as the notion that the

> g-field is entirely determined by the masses
> of bodies. Mass and energy, according to the
> conclusions of Special Relativity, are essentially
> equivalent; formally energy is described by the
> symmetric energy tensor, i.e. the g-field is
> defined and determined by the energy-matter tensor.
> (Einstein, 1918)

Unfortunately it is immediately clear that this form of Mach's Principle is not compatible with the general theory. For, in the case where there is no matter at all in the space-time, one nevertheless obtains from the field equations at least one acceptable space-time structure, namely the space-time of special relativity (see also Ozsvath & Schücking, 1962). Einstein's reaction to this was to endeavour to modify the field equations, hopefully to render such solutions impossible. That is, he tried to find an equation to which there is no solution at all, hence no space-time, when the stress-energy content is put equal to zero. If this were possible then not only would the geometry of space-time and the motion of freely falling bodies depend on the matter in the world, but the very existence of space-time would depend on matter.

Einstein's modification of his original field equations to this end consisted in the addition of an extra term, the so-called cosmological term, of the form $\Lambda g_{\mu\upsilon}$, to the left hand side of the equations. Λ is called the cosmological constant, and is a free parameter of the modified theory to be determined by experiment. The two versions of the theory, both with and without the cosmological term, are referred to as the general theory of relativity, specific reference to the cosmological term being added if necessary. There are two objections to the inclusion of a non-zero cosmological constant which we shall explain in detail below. The first is that it is against the spirit of Einstein's interpretation of Mach's ideas, and the second that it fails to achieve the desired objective. Of course, if a non-zero value for Λ were ever discovered experimentally these objections would have to be overcome. Observations of the perihelion precession of Mercury to better than *10%* imply that $\Lambda < 10^{-34}$ cm^{-2}. The absence of cosmological effects over the visible Universe indicate $\Lambda < 10^{-54}$ cm^{-2}. Of course, no experiment could ever reveal $\Lambda = 0$ exactly.

The introduction of the cosmological term violates the spirit of relativity since, at least in the first instance, it attributes to space a property which is independent of matter, namely the possession of a non-zero cosmological constant. Thus it is a property of space which acts on matter but is not itself acted upon. In this sense it is an absolute attribute of space-time. This objection could be overcome if it could be shown that the cosmological term is a result of material processes in a curved space-time. For example, attempts have been made to describe the cosmological term as representing a contribution to the vacuum energy of space-time due to (virtual) quantum processes (Zel'dovich, 1967). Such theories should, of course, predict the value of Λ. From this point of view the term really arises on the right hand side of the field equations, representing the quantum nature of matter.

The second objection, that the introduction of the cosmological term fails to achieve its objective, is not, of course, an argument against its inclusion, but shows conclusively that the general theory of relativity does not justify Einstein's form of Mach's Principle. A solution of the modified field equations with zero matter content was found by de-Sitter (1917). Explicitly the metric

$$ds^2 = \frac{1}{K}\left(-d\theta^2 + \cosh^2\theta\,(d\chi_1{}^2 + \sin^2\chi_1\,(d\chi_2{}^2 + \sin^2\chi_2 d\chi_3{}^3))\right)$$

satisfies the modified empty space field equations. The geometry of the de-Sitter solution can be described as the surface of a four-dimensional hyperboloid embedded in a five-dimensional Minkowski space. It has a certain historical role, even apart from the Mach problem, since it was thought at one time that the divergence of test particles moving on the geodesics of the de-Sitter metric might bear some relation to the expansion of the Universe. And, in fact, part of the de-Sitter hyperboloid is the space-time of the Steady State cosmological theory. However, as soon as Einstein became aware of de-Sitter's solution, he recommended that the cosmological term be omitted, describing its proposed existence as the 'biggest blunder of my life'. The full history of the Λ-term is chronicled by North (1965).

It is now known that there are many solutions of the field equations, with or without the cosmological term, representing the geometry of space-times containing no matter. These solutions may be thought of as representing universes consisting of gravitational energy only, with no material sources. Whether any meaning could be attached to such space-times is not relevant; general relativity admits their existence, so does not automatically satisfy Einstein's statement of Mach's Principle.

§9.2   *Mach's Principle and asymptotically flat space-times*

The space-time metric determines the motion of bodies under no non-gravitational forces through the geodesic equations. If the metric were determined by the material content of the World, it follows that the motion of massive bodies would also be so determined, and in this sense we could say that the inertial properties of matter arise from interaction with the rest of the Universe. Thus, Einstein gives an alternative formulation of Mach's Principle:

> In a consistent theory of gravitation there can
> be no inertia relative to "space", but only an
> inertia of masses relative to one another. If,
> therefore, I have a mass at a sufficient distance
> from all other masses in the Universe, its inertia
> must fall to zero.   (Einstein, 1917)

Again, this fails to be satisfied automatically in general relativity. For the simplest example, we can turn to the case of a spherically symmetric distribution of matter of finite extent in an otherwise empty space, the metrical properties of which are to be determined by solving the Einstein field equations (with $\Lambda = 0$). This solution provides a mathematical model for the space-time of the solar system dominated by the Sun in which the

140

planets move as test bodies. It was first obtained exactly by Schwarzschild (1916), although Einstein (1915) had earlier used an approximate form in his discussion of the perihelion advance of the planet Mercury. The Schwarzschild solution is appropriate to the exterior vacuum space-time independently of the composition of the central mass, and is indeed valid even if the matter undergoes time dependent oscillations provided only that it remains spherical (Birkoff, 1923). There is, of course, a precise analogue of this result in Newtonian theory.

The form of the metric, in the coordinates adopted by Schwarzschild, is

$$ds^2 = -c^2 \left(1 - \frac{2GM}{rc^2}\right) dt^2 + \left(1 - \frac{2GM}{rc^2}\right)^{-1} dr^2 + r^2(d\theta^2 + sin^2\theta d\phi^2) \ .$$

$$(9.2.1)$$

This solution is obtained by imposing the condition that as $r \to \infty$ the metric coefficients should converge to those of Minkowski space-time in an inertial frame. Space-times satisfying this condition, a large number of which are known explicitly, are called asymptotically flat. Indeed the significance of the constant M as the total mass of the central body is demonstrated by observing that the trajectories of freely falling bodies at large values of r are approximately what one would expect in Newtonian theory if the velocities are sufficiently small that special relativistic effects can be neglected. It is not necessary to make this assumption concerning the boundary conditions, and indeed, if one wanted to model the solar system with greater accuracy, one could take into account the acceleration of the Sun relative to the kinematical inertial frame of the 'fixed stars' by requiring an asymptotic approach to the Minkowski metric in coordinates in which the central mass has an appropriate motion. And one could model a geocentric Universe by a choice of boundary conditions appropriate to the motion of the Sun about the Earth. Here, then, we see explicitly the vestiges of Newtonian absolute space in the role of the boundary conditions to be imposed on solutions of the field equations.

It is now clear that the form of Mach's Principle proposed by Einstein in the quotation at the beginning of this section is not embodied in the general theory of relativity. For in the Schwarzschild solution at an arbitrary distance from the central mass the inertial properties of a test body are precisely the same as in special relativity: the acceleration of the body produces an inertial force even in the absence of material sources; the acceleration of the body occurs relative to 'space'. The Schwarzschild solution may not be appropriate to our Universe -

141

except as a local model for the solar system - since this is presumably not empty of matter at sufficiently large distances. However, the point is again that if general relativity admits such a possibility, it cannot satisfy this form of Mach's Principle.

§9.3 *Causal problems*

The metric of a space-time determines its causal structure. Locally this reduces to the light cone structure of special relativity dividing events into those in the observer's past and future and those which are inaccessible at a given moment. Globally the effect of gravity on the propagation of light may mean that the local pictures are built into a counter-intuitive global picture which, moreover, may raise problems for the Machian approach. We can illustrate this in the case of the Schwarzschild solution.

Consider what happens to the metric form (9.2.1) as r approaches the so-called Schwarzschild radius, $r_g = 2GM/c^2$. It would appear that the time interval between events with arbitrarily different t-labels approaches zero, and the spatial separation between points with arbitrary r-labels becomes infinite. There are four possible solutions that could be proposed:

(i)   The problem is an artifact of the assumption of exact spherical symmetry. Since exact symmetry cannot arise in nature because a statistical system will always exhibit fluctuations, the problem is of no physical relevance.

(ii) It is physically impossible that r should approach $2GM/c^2$ since this says that matter must be compressed to a small volume, which for real matter might initiate processes which prevent further contraction. In this case the investigation of the problem might be a mathematical exercise without physical significance.

(iii) The boundary $r = r_g$ represents an edge to space-time, a physical boundary beyond which one may not travel. For example, as a body approaches $r = r_g$ it might experience gravitational forces which tend to infinite strength and ultimately destroy it.

(iv) The description of the space-time in Schwarzschild coordinates may be inappropriate at this point. For example the spherical polar coordinates θ, ϕ become inappropriate at $\theta = 0$ since we see from equation (9.1) that the distance between points with arbitrarily different ϕ labels is zero at the pole $\theta = 0$. Nevertheless, the pole is a perfectly ordinary point of the sphere.

142

Solution (i) is ruled out by uniqueness theorems for the end-points of the collapse of massive objects. Any initial asymmetries are radiated away and, if the initial object is non-rotating, the final configuration approaches exact spherical symmetry (see Carter, 1979, for a review of the extent to which this has been rigorously proved).

Solution (ii) is unacceptable since the densities involved need not be large. An object the size of the galaxy composed of water would satisfy the condition that its radius would be less than its mass x $2G/c^2$, and we know that in principle no exotic physical process could prevent such a structure from forming. In fact, if we abide by known physics, including general relativity, objects of sufficient mass which have reached a certain density are actually aided in their collapse to a configuration of radius smaller than r_g by any processes which might have been thought to prevent collapse. This is because, beyond a certain point, any pressure force preventing collapse effectively contributes to a greater degree to the gravitational mass of the body causing the collapse. We might therefore reasonably expect to find objects which are 'smaller than' r_g.

The possibility (iii) appears unlikely since one can compute the magnitude of the expected gravitational forces and these are not infinitely large. In principle one could still have an edge to space-time; this is ruled out by the explicit construction of an alternative coordinate system showing (iv) to be correct.

A particular choice of coordinates which allows the extension of the space-time beyond $r = r_g$ to be demonstrated was given by Kruskal (1960). Define new coordinates u and v for $r > 2m$ by

$$u = (r/2m - 1)^{\frac{1}{2}} e^{r/4m} \cosh(tc/4m)$$

$$v = (r/2m - 1)^{\frac{1}{2}} e^{r/4m} \sinh(ct/4m)$$

where $m = GM/c^2$. Clearly we have $u > |v|$. The Schwarzschild metric assumes the form

$$ds^2 = \frac{32m^3}{r} e^{-r/2m} (-dv^2 + du^2) + r^2 (d\theta^2 + \sin^2\theta d\phi^2) \qquad (9.3.1)$$

with r as a function of u and v given implicitly by

$$(r/2m - 1) e^{r/2m} = u^2 - v^2.$$

In this form there is manifestly no singularity at $r = 2m$ and
the metric can be extended beyond the quadrant $u > |v|$. The
surface $r = 2m$ has no significance locally, and an observer
would notice no effect on passing through the surface. However,
the surface does have a global significance. To see this, consider a space-time diagram appropriate to the $(u - v)$ plane of
the metric (9.3.1) (Fig. 9.1).

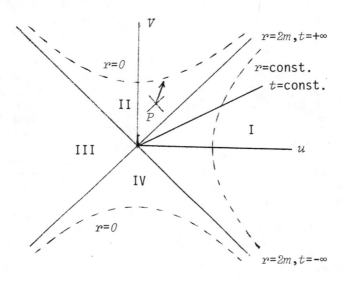

Fig. 9.1. Kruskal coordinates (u,v). The whole $(r\text{-}t)$ plane of
the Schwarzschild solution is represented by quadrant
I. Any path to the future of a point in II, such as
P, hits the singularity at $r = 0$.

From the expression for r as a function of u and v we see
that the line $r = 2m$ (in the $r\text{-}t$ plane) corresponds to $u = \pm v$,
and therefore $t \equiv 4m \ tanh^{-1} \ v/u = \pm\infty$. At $r = 0$ there is a real
singularity at which not only the metric coefficients but also
the gravitational tidal forces become infinite. Continuation
beyond $r = 0$ is mathematically possible but physically irrelevant.
The $r = 0$ singularity appears in the $u\text{-}v$ plane as the *two* hyper-
bolae $u^2 - v^2 = -1$. There are two singularities and two asymptotically flat regions (I and III) in each of which the local
physics may be represented by a metric of the Schwarzschild form.
These two asymptotically flat universes are joined by a 'throat'
or 'bridge' corresponding to regions II and IV, in which the geo-

144

metry is time dependent; in II a singularity is reached after a finite time (see, e.g. Misner *et al*. 1973).

The Kruskal coordinates are chosen such that the local light cones are represented exactly as in special relativity by lines at 45° (using units in which length is measured in light seconds, so $c = 1$). It is then immediately obvious that the lines $r = 2m$ are the paths of light rays which travel out to infinity, and that any light emitted in region II falls into the singularity. Once a body has entered region II it cannot escape, since all future time-like paths from a point in II hit the singularity and do not get to infinity. The surface $r = 2m$ is called an *event horizon* to express this global significance; it marks the boundary of the region of space-time from which light, and hence, matter, can escape to infinity.

The surface $r = 2m$, $t = +\infty$ constitutes the space-time boundary of a *black hole*. In a similar way one can see that nothing can get into region IV from region I, whereas matter may be arbitrarily ejected from region IV into region I. For this reason the surface $r = 2m$, $t = -\infty$ is referred to as the boundary of a *white hole*. To represent the physical collapse of a star to the black hole configuration one substitutes a region of the space-time diagram representing the interior of the star. The effect of gravity on light, and hence on the causal structure, is clearly seen in a representation of the collapse in Eddington-Finkelstein coordinates (Fig. 9.2).

The continuation of the space-time beyond the surface $r = 2m$, $t = +\infty$ is not uniquely determined by the distribution of matter and the geometry in the exterior region I. The extension we have given is the smoothest possible one (analytic) but any number of other possibilities exist in principle. This is important for the implementation of Mach's Principle, since this requires that the global properties of space-time be uniquely determined in some way by the matter distribution. One might argue that the problem here is not serious for two reasons. First, the exterior geometry is uniquely determined and the non-uniqueness is enclosed in a region of space-time about which an external observer can have no knowledge. The counter argument is clearly that there is no necessity for the more dedicately curious observer to remain on the outside of the black hole. Second, the space-time geometry is at least uniquely determined by a knowledge of the state of the system everywhere at any one time, i.e. on any spatial surface. We shall make this more precise below. It is, however, possible to construct space-times for which this is not the case; for which the future is not determined by the data at one time, but for which 'initial' data must be given anew to determine the con-

145

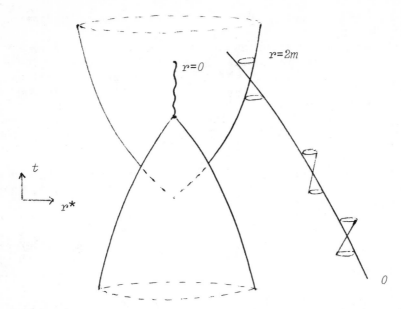

Fig. 9.2. A collapsing star in Eddington-Finkelstein coordinates
$(t,r^{\star})$. The tilting of the light cones along the path
of an observer O falling into the black hole shows the
effect of gravity on light.

tinuation of the evolution of the system beyond some point. Such
a space-time is that which represents the end-point of the collapse of a rotating star, namely the space-time of a rotating black
hole. This will raise also some problems of causality.

   The space-time of a rotating black hole has a metric given
locally by the Kerr (1963) solution,

$$ds^2 = \rho^2(\Delta^{-1}dr^2+d\theta^2) + (r^2+a^2)sin^2\theta d\phi^2-dt^2+2mr^{-2}(asin^2\theta d\phi-dt)^2$$

where $\rho^2(r,\theta) \equiv r^2+a^2cos^2\theta$ and $\Delta(r) = r^2-2mr+a^2$, and again we use
units with $c = 1$. The constant $M = c^2m/G$ can be interpreted as
the mass of the hole and Ma as its angular momentum measured at
infinity. For $a^2 < m^2$ we find $\Delta(r) = 0$ at $r = r_+ \equiv m+(m^2-a^2)^{\frac{1}{2}}$,
and these are coordinate singularities in the Kerr metric analog-
ous to the singularity at $r = 2m$ in the Schwarzschild solution.
There is no local pathology of the space-time at $r = r_\pm$. By suitable coordinate transformations one can extend the space-time
through these surfaces (Fig. 9.3). The extension consists of
three distinct regions. The space-time outside $r = r_+$ is asympto-

146

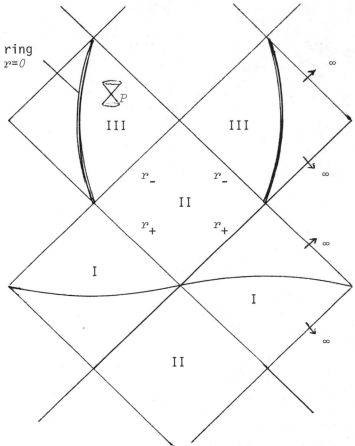

ring
$r=0$

III

III

∞

r_-

r_-

∞

II

r_+

r_+

∞

I

I

∞

II

Fig. 9.3.   The maximal extention of the Kerr solution (with $a^2 <$ m^2).   The pattern of regions I, II, III repeats end- lessly.   Regions I and III extend to asymptotically flat past and future infinity.

tically flat, and $r = r_+$ is the event horizon which bounds the region from which bodies may escape to infinity (I). Region III is $r < r_-$ and contains a true singularity with infinite tidal forces at $r = 0$. However, the singularity has the form of a ring; continuation of the space-time to negative values of r is possible through the disc which spans the ring and one passes into another asymptotically flat 'universe'. Region II is $r_- <$ $r < r_+$. A further extension of the space-time is then possible

147

in which the three regions become components which can be joined together as shown in Fig. 9.3.

Now a null trajectory to the past of point P in region III can be continued into the singularity or to past infinity without crossing the surface S. Conversely, this means that information, in the form of electromagnetic radiation, say, can be received at P without being registered on S. Consequently, knowledge of the state of the system on S is not sufficient to allow a prediction of the state at P, some time in the future of S. The surfaces $r = r_-$ form the boundary of the region of space-time determined by data on S. In such a space-time we should have to side with Newton in the discussion with Leibniz, not because our laws would fail to predict correctly, but because they would fail to predict at all.

Strictly speaking one should investigate the stability of this space-time structure to small perturbations. If it is not stable then no amount of electromagnetic radiation could be introduced without qualitatively altering the space-time picture, so no information could be propagated to P. It would appear that the structure is indeed unstable (Simpson and Penrose, 1973). Nevertheless, in a universe satisfying Mach's Principle in some form, we should still wish to rule out the possibility that there are regions of space-time the nature of which cannot be determined by the matter distribution and geometry at some earlier time. This does not, of course, mean that we should wish to disallow the formation of rotating black holes, but we should require that the region exterior to black holes have a predictable evolution. This requirement will be made precise in the assumption of 'global hyperbolicity' for the space-times with which we shall deal, a notion which will be discussed in Section 12.3.

We have not exhausted the violations of our intuitive notions of causality which appear in the Kerr solution. For in the vicinity of the ring singularity there are time-like paths which one can follow from some given space-time point, P, into the future until one arrives back at the point P! If such closed time-like paths were to exist in the real universe, certain unnatural limitations to possible actions would ensue. For suppose P were some short time prior to the publication of Einstein's 1916 papers on the general theory of relativity. Then the reader might wish to achieve a certain notoriety by using his presently acquired expertise in the field and return to P to publish his own account predating Einstein. Since this would present us with two coexisting and incompatible futures the resolution is presumably that, in such a world, neither Einstein nor anyone else could have discovered the theory of relativity. The argument may

148

clearly be extended to any other activity.

Such causality violations must clearly be ruled out in a physically acceptable model of the world. Solutions of Einstein's equations which exhibit them are of no physical relevance. In the case of the Kerr black hole one usually argues that it is only the exterior solution that is relevant, the analytic extension to the interior being to a certain extent arbitrary. The singularity and the violations of causality are hidden from the outside world by the event horizon. Whether it is possible to construct a singularity, which is not so hidden, from a normal state of matter without violating known physics is an unsolved problem in general relativity. The assertion that unhidden or 'naked' singularities do not exist in called the *cosmic censorship hypothesis* (Penrose, 1974).

The non-existence of closed time-like or null paths is not yet a sufficient condition for a respectable solution of Einstein's equations. It is reasonable to exclude also time-like or null paths which return arbitrarily close to their point of origin, or which pass arbitrarily close to curves which themselves return close to their point of origin, and so on. Carter (1971) has shown that there is a hierarchy of such distinct causality conditions. Since any physical system is subject to fluctuations one would like to ensure that the attendant perturbations of the metric do not produce space-times which violate any of this hierarchy of conditions. This requirement is the condition of *stable causality*. It can be shown that a consequence of this condition is the existence of a function which increases along every future-directed time-like or null trajectory and which may therefore be considered to play the role of a cosmic time (Hawking 1968, 1971). For the development of the theory in Chapters 10 and 12 we shall require not merely the existence of a cosmic time, but the stronger condition of global hyperbolicity, that the state of the World at any one time should determine its future evolution.

§9.4   *Machian effects in general relativity*

Even if Einstein's version of Mach's ideas are not fully incorporated in the final theory, one can nevertheless look for what might be called Machian effects. For example, one might hope to demonstrate explicitly from the point of view of an observer at rest on the Earth, for whom the stars are in rotational motion, how the gravitational forces exerted by the distant matter in the Universe cause the deformation of the Earth. Note that we know this must be equivalent to the picture in which the Earth rotates and the heavens remain at rest from the covariance

149

of the theory, and this does not contradict anything we have said above concerning the absolute nature of acceleration in general relativity. The distinction is this: once we know how to get a correct answer in one frame of reference, we are guaranteed to obtain the correct result in any other frame; the element of absoluteness enters relativity in the extra assumptions, concerning boundary conditions, which we have to make in order to predict correctly the gravitational-inertial forces in any one frame.

The first attempt in this direction was made by Einstein himself. As early as 1913 on the basis of general relativity as then conceived, Einstein wrote to Mach:

> For it necessarily turns out that inertia
> originates in a kind of interaction between
> bodies, quite in the sense of your considerations
> on Newton's pail experiment.
> (Quoted by Misner et al. 1973, p. 544)

The consideration to which Einstein is referring is Mach's claim that the outcome of the bucket experiment might well be different if performed with a bucket 'several leagues thick'. Einstein believed that certain effects should follow in a theory which incorporated Mach's ideas:

(i)   the inertia of a body must increase with the agglomeration of mass in its neighbourhood.

(ii)   a body should experience an acceleration if nearby bodies are accelerated. The accelerating force should be in the same direction as the acceleration of the bodies.

(iii)   a rotating hollow body should generate inside itself a Coriolis field which deflects moving bodies in the sense of the rotation, and a centrifugal field as well (Einstein, 1955).

We shall discuss each of these in turn.

Einstein tried to demonstrate the first effect by means of an approximate treatment in which the metric of the space-time is taken to be that of Minkowski space, and deviations from straight-line free falls are attributed to the existence of forces, or, in this particular case, to an increase in the inertial mass of a test body. The extraction of physical information from approximation schemes in general relativity is now known to be particularly difficult. This arises essentially because the metric controls not only the physics, but also the arbitrary meaning of the coordinates. In an approximate treatment it is very difficult to separate a physically significant perturbation from an arbitrary change in the meaning of the coordinates. For suppose we make a small transformation of coordinates. In the new coordinates the metric has the approximate Minkowski form for a new set of observers, namely those moving on the new coordinate time lines. Thus a revised choice of Minkowski space-time background is accompanied by a new choice of the observers who are to be regarded as inertial observers in the absence of the gravitational perturbation. This is equivalent to an attempt to specify an inertial frame in the presence of gravity - a frame in which to apply a special relativistic interpretation of gravitational forces. We saw that the impossibility of a special relativistic theory of gravity and the development of a general relativistic metric theory arises from the fact that, because of the principle of equivalence, such a choice of inertial frame cannot be made in an unambiguous way. Thus, without due care, any results derived on the basis of such a choice may be merely a reflection of this arbitrariness and without physical significance. This is the fate of Einstein's demonstration (Brans, 1962). In general relativity inertial mass is an intrinsic property of bodies and is uninfluenced by external entities. The distant matter in the Universe does not generate inertial mass but inertial forces, which are masses multiplied by accelerations.

The second effect claims to show explicitly this induction of acceleration in a body by accelerating masses. The effect certainly occurs and it is perhaps helpful to interpret it in this way. In particular it gives insight into how the theory differs from Newtonian gravity by incorporating gravitational forces in addition to the inverse square law. Such an interpretation is nevertheless not in accordance with the spirit of the theory and is not essential. One could alternatively say that the presence of accelerating matter changes the freely falling trajectories in its neighbourhood, through the generation of space-time curvature, and consequently changes the motion of freely falling matter (see Zel'dovitch and Novikov 1967, pp. 586-587).

Similar remarks apply to the third effect since this is only another aspect of (ii). Nevertheless, this problem has a long and apparently unending history. The effect was discovered by Thirring (1918, 1921) in an investigation of the metric inside a thin rotating spherical shell, and by Lens and Thirring (1918) for the rotation of a solid spherical mass. Essentially one writes the metric $g_{\mu\nu}$ as a sum $g_{\mu\nu} = n_{\mu\nu} + h_{\mu\nu}$ representing the small departure, $h_{\mu\nu}$, from Minkowski space-time $n_{\mu\nu}$, and neglects quantities of order $h_{\mu\nu}^2$ and higher. This yields the weak field approximation to the field equations of general relativity, and the results are supposed to be interpreted in terms of gravita-

151

tional forces in flat space-time. From this point of view, in the Lens-Thirring effect the rotating body drags the local Lorentz frames round with it. Thus a projectile aimed at the centre of a rotating body will be dragged round by the body and will not move radially (Fig. 9.4). Nevertheless the projectile does not acquire

Fig. 9.4. A projectile aimed at the centre of a rotating body is dragged round by the (non-radial) gravitational field.

angular momentum, since the local Lorentz frame falls freely with the body; only rotation with respect to this frame, not the frame at rest at infinity, generates inertial forces. This dragging effect can be illustrated in an extreme case in the Kerr metric of a rotating black hole. Outside the horizon $r = r_+$ there is a surface given by $r = m + (m^2 - a^2 \cos^2\theta)^{\frac{1}{2}}$ called the 'stationary limit'. Between the horizon and the stationary limit the dragging effect of the hole is so strong that no body can remain at rest with respect to infinity, no matter how powerful a thrust it can develop to cause its motion to depart from free fall.

There is an inconsistency in the work of Thirring; he assumed the particles of the shell to move on circular orbits but took no account of the stresses needed to make them do so. For without these stresses the particles of the rotating shell will simply fly apart. This inconsistency was resolved by Bass and Pirani (1955). They found a dragging of inertial frames with angular velocity $4/3$ $GM\omega/r$ for a shell of mass M and radius r rotating with angular velocity ω; they also pointed out that the ω^2-terms, which Thirring had interpreted as centrifugal forces, could be regarded as due to the special relativistic variation of density between the poles and equator of the shell arising from variations in velocity. The interpretation of the ω^2 terms has also been discussed by Lausberg (1969, 1971) and by Teyssandier (1972a, b). Lausberg proposed a distinction between terms of order $\gamma\omega^2$, which represent the kinetic energy of rotation equivalent to a relativistic increase in mass, and terms of order $\gamma^2\omega^2$ which represent true centrifugal terms. The analysis has been carried out by Teyssandier.

Brill and Cohen (1966; Cohen and Brill, 1968) treat the problem of rotating shells as a perturbation of the Schwarzschild geometry, rather than of flat space-time. They show that, for strong gravitational fields, the contribution to the dragging effect from each of several shells is altered by the presence of the others and cannot be added in any simple way. For a slowly rotating ball of pressure free matter in an asymptotically flat universe, Brill and Cohen find that as the radius of the ball approaches the Schwarzschild radius, $r_g = 2GM/c^2$, the interior geometry approaches that of a closed universe (Section 9.5); in the limit of complete closure, the rotation rate of the ball coincides with that of the local inertial frame. (For a review, see Heller 1975).

§9.5   *Mach's Principle and cosmology*

One should perhaps take the view that Mach's Principle, in whatever precise form, relates local physics to the global structure of the Universe and as such cannot be investigated by means of solutions of the field equations appropriate only to local physics. Of course, it can be argued that the Schwarzschild solution, say, is a perfectly acceptable representation of the space-time of a universe containing a single spherical body and, since the Universe could have been like this, it follows that general relativity does not embody Mach's Principle. This argument is quite correct, but it might nevertheless be important to see the extent to which universes more like our own with, for example, matter everywhere, might encompass Mach's ideas. For this reason Einstein was led to initiate the modern scientific study of cosmology with the construction of a model of the Universe.

In the construction of such a cosmological model we are clearly not concerned with small scale details. The first step therefore is to decide which details shall count as small scale, and which features shall be modelled in detail. This is closely related to the assumption of a cosmological principle. We wish to take some average over certain regions of, for example, the density of matter, and let that average density represent the material content of the region. In this way we eliminate all details such as the motion of planets and the types of stars. It is, however, not at all clear that such averages are meaningful. Consider for example a hierarchical universe consisting of galaxies and clusters of clusters of galaxies and so on. One can imagine these as arranged in such a way that at each stage one determines an average density of matter which is smaller than that determined by the previous scale of clustering. To do this one has only to make the distances between successive orders of clustering increase sufficiently rapidly with order. In this situation no meaning can be attributed to the average density of matter, and one would have to deal with average densities defined for each scale.

If we could see the whole Universe questions such as this could be decided by observation. However, in practice we certainly cannot see the whole Universe, and, even in principle, we believe it may be impossible to do so (see below). It is therefore necessary to establish a cosmological principle making assertions about the Universe as a whole which can only be checked indirectly by the predictions that can be extracted from models constructed on the basis of the adopted principle. If we adopt the hierarchical principle suggested above it appears to be impossible to construct models in conformity with observations of certain radio radiation received on Earth from distant parts of the Universe. This is the microwave background radiation which we shall discuss more fully in Section 12.6. Thus it would appear that by this indirect means we are able to rule out the hierarchical cosmological principle.

Direct observation of that part of the Universe which is accessible to us seems to indicate a distribution of matter which is isotropic when averaged over distances large enough to contain several clusters of galaxies. There are two possible explanations. Either we are at the centre of a spherical distribution, or the distribution is homogeneous and would therefore exhibit the same isotropy for any observer. Since it is unlikely that the Earth occupies such a privileged position, and since this would go against the spirit of the Copernican revolution, it is natural to assume that the distribution of matter in space is uniform on a sufficiently large scale, and to adopt this as a cosmological principle.

The isotropy will not be apparent to all observers, since the motion of an observer will produce a Doppler shift to higher frequencies of light received from the direction of motion. Therefore the assumption of isotropy of the spatial distribution singles out a privileged class of observers which we take to be those falling freely with clusters of galaxies. Other observers

154

in relative motion do not see the isotropy directly. The atomic clocks of the privileged observers measure a privileged cosmic time t. Space sections at constant time t are privileged spaces of simultaneous events, and it is just these space sections which are homogeneous in general. In the particular case when nothing in the universe changes with time it is possible to find other homogeneous space sections. Einstein's original model of the Universe was of this form. However, the study of cosmology seems to have revealed that the Universe is evolving. In particular it is expanding, in the sense that clusters of galaxies are moving apart; and the rate of expansion is slowing down. Consequently conditions change from one space section to the next and the slicing up of the space-time into homogeneous space sections can be carried out in one way only.

The metric of a space-time satisfying these conditions of homogeneity and isotropy is of the form

$$ds^2 = - dt^2 + R^2(t) \ (dr^2 + r^2(d\theta^2 + sin^2\theta d\phi^2)) \qquad (9.5.1)$$

where $k = 0$ or ± 1, and the function $R(t)$ is determined by Einstein's field equations. This form of the metric was discovered by Friedman (1922, 1924) and Lemaitre (1927, 1933) and investigated by Robertson (1935, 1936) and Walker (1936). The three possible values for k determine the three possible geometries of the space sections t = constant. For $k = 0$ these spaces are flat Euclidean spaces; for $k = +1$ they are three dimensional spheres (i.e. the closed surfaces of spheres in four-dimensional space); for $k = -1$ they are three dimensional hyperboloids of constant negative curvature. If the cosmological constant is zero the value of k is determined by the density of matter in space; k is positive if the density is greater than some critical value. The Robertson-Walker or Friedman metric (9.5.1) describes a space-time expanding uniformly, with distances between freely-falling paths proportional to the 'scale factor' $R(t)$. In the case $k = +1$ the expansion is reversed after a finite time and the Universe recollapses. In the other cases the expansion has a beginning at a finite time in the past, the so-called 'big bang', but will continue expanding forever.

The space-time diagram for these models (Fig. 9.5) shows the world lines of material particles cutting orthogonally the three dimensional spaces of constant cosmic time, since the spatial velocity (or 'proper motion') of these particles is zero. The particles may be thought of as mathematical elements of a continuous smooth fluid representing the average density of

155

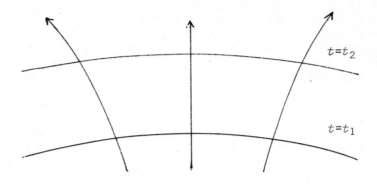

Fig. 9.5. The world lines of uniformly receding particles are orthogonal to the homogeneous three-spaces of constant cosmic time, t, in the Friedman-Robertson-Walker models.

matter, or as the clusters of galaxies viewed on a scale on which they have no internal structure, i.e. on which they can be regarded as points. The clusters themselves do not expand since they are bound by their own gravitational fields. The expansion of the Universe is manifested in the increasing length of the space-like vector connecting neighbouring particles. This shows how the Universe can be spatially homogeneous while every particle appears to be expanding away from a given observer.

An important feature of these models is that for a large range of physically realistic choices of $R(t)$ there exist *horizons* from beyond which no information can have been received at a given time. As the Universe expands, more and more of it becomes accessible to observation by a given observer. Nevertheless, at any finite time in such a Universe, the cosmological principle, which was assumed in order that the model might be constructed, cannot be checked, even in principle, by direct observation.

In these model universes an observer in free fall sees distant matter expanding away from him without any shearing or rotational motion. In his local Lorentz frame the distant stars are observed to have no transverse motion, so one can say that Mach's requirement of the agreement of the local dynamical and

156

kinematical inertial frames is satisfied. The question arises as to whether this result has again been imposed by a special choice of solution to the field equations, or whether it somehow arises automatically from consideration of the equations in conjunction with the cosmological principle. This question will be answered by finding solutions in which relative rotation of the two frames is present.

The question can be raised also as to whether these models satisfy the requirement that the space-time geometry be completely determined by matter. It is necessary to establish precisely what is meant by 'determination' in this context. For example, Einstein (1917) conjectured that spatially closed universe models, such as the $k = +1$ Robertson-Walker model, would satisfy his Mach's Principle because the absence of a boundary implies there is no need for boundary conditions. On the other hand priority has been claimed in this respect for the $k = 0$ Robertson-Walker models (with $\Lambda = 0$). These provide a relation between the gravitational constant, G, and the density of matter in the universe of the form $G = 3H^2/8\pi\rho$ where H, Hubble's constant, is the ratio of the velocity of recession of a distant galaxy to its distance from us. Consequently H^{-1} is a measure of the time since the galaxies would have been together at a point - an exact measure if the expansion were uniform - hence a measure of the age of the Universe. The size, R, of the visible Universe can be estimated as the distance over which light could have reached us in the time H^{-1}, so $R = cH^{-1}$, and we can write

$$\frac{1}{G} \sim \frac{c^2}{R} \quad \frac{4}{3}\pi\rho R^3 \sim \frac{c^2 M}{R}$$

where M is the mass of the visible Universe. In this sense the strength of gravity in these models is apparently determined by the matter content. We shall take up the discussion of the meaning of determination again in Section 12.3, where we shall propose a solution for which neither of the above considerations are valid.

Let us then return to the question of alternative solutions of the field equations. In order to obtain such solutions we must weaken the statement of the cosmological principle, since the Robertson-Walker models exhaust the class of spatially homogeneous isotropic models in general relativity. It is not possible to drop the homogeneity requirement without at the same time allowing anisotropy, since inhomogeneity implies spatial gradients and hence privileged spatial directions. However, we can allow models which are spatially anisotropic but homogeneous; at each point there will be privileged directions but these will be the same from point to point.

We therefore seek cosmological models containing spatial sections of constant cosmic time on which the geometry is homogeneous but anisotropic. The space-like vector connecting neighbouring particles of matter (or clusters of galaxies) describes the relative motion of those particles. More general motions will now be possible in addition to a uniform expansion. In general the neighbouring particles may separate at different rates in different directions. This will give rise to a shearing motion, in which a circle of particles evolves into an ellipse, and a rotational motion, in which the axes of the ellipse rotate, in addition to an overall expansion. It is also necessary to take account of the possibility that the material particles have a proper motion with respect to observers for whom the geometry of space is homogeneous, and of the possibility that external forces, such as pressure forces, may cause the motion of the material particles to deviate from geodesics. In the former case we refer to 'tilted' models (see King and Ellis, 1973). In these, just as in models with rotation, the world lines of particles do not cut the surfaces of constant cosmic time orthogonally. In the latter case the four-velocity of matter has a component representing the acceleration relative to geodesics. The change in the velocity of matter from point to point consists of the sum of all these contributions.

Clearly we can look for simplified models in which only some of these effects are present. The simplest case is that in which in addition to overall expansion there is only shear present. The various possibilities are classified according to the remaining symmetry of the spatial sections. As usual these symmetries can be described in terms of the transformations which can be carried out on the space without altering anything but the labelling of the points. This can be thought of from an active point of view as the ways in which the space can be slid over itself. In the homogeneous isotropic models the symmetries can be described in terms of groups of transformations involving six parameters. The simplest example is the $k = 0$ case in which the space sections are ordinary Euclidean three dimensional space. Here we may translate the space in one of three independent directions (3 parameters) and we may rotate through any angle (1 parameter) about any axis, the direction of which is described by giving two angles (2 parameters). Technically the transformation

$$\underset{\sim}{x} \to \underset{\sim}{x}' = R\underset{\sim}{x} + \underset{\sim}{a} \quad ,$$

158

where R is a matrix describing a rotation and $\underset{\sim}{a}$ is a vector representing the translation, does not change the Euclidean metric

$$ds^2 \ = \ dx^2 + dy^2 + dz^2 \quad .$$

The geometry is not defined by the number of parameters alone, since the $k = \pm 1$ cases with spherical or hyperbolic geometries admit groups of transformations with the same number (6) of parameters. To see how the difference arises we consider the effect of performing two transformations in succession. For simplicity take a two-dimensional example, namely the Euclidean plane and the two dimensional surface of a sphere. In the plane, two translations may be carried out in either one of two possible orders with the same result. On the sphere one can easily see that the result depends upon the order in which the 'translations' are made.

Thus, the classification of possible spatially homogeneous anisotropic models with shear only, amounts to first of all an investigation of the number of parameters the symmetry transformations may have (4 or 3 are both possible), and subsequently a classification of the different ways in which pairs of transformations may combine when taken in different orders. This will characterise the types of transformations and hence the possible symmetries. The Einstein equation must then be solved to show how the spatial geometry evolves in time.

The appropriate classification was first investigated by Bianchi (1918) improved (1920) and applied to cosmological problems by Estabrook et al. (1968) and MacCallum and Ellis (1970). In general the Einstein equations cannot be solved explicitly to yield a form for the metric in terms of elementary functions, although many explicit exact solutions are known. As an example consider the metric

$$ds^2 \ = \ - dt^2 + X^2(t)dx^2 + Y^2(t)dy^2 + Z^2(t)dz^2 \quad .$$

Clearly this respresents a model in which the rate of the expansion in three orthogonal directions is controlled by the functions X, Y, Z. The Einstein equations yield particular functions for X, Y, Z depending on the assumed matter content. For example, if the space-time is assumed to be empty we find the *Kasner* solutions with $X(t) = t^{P_1}$, $Y(t) = t^{P_2}$, $Z(t) = t^{P_3}$ and $P_1 + P_2 + P_3 = 1 = P_1{}^2 + P_2{}^2 + P_3{}^3$.

Bondi (1960) has suggested that shearing motions in the Universe would be a non-Machian feature since the shear would mimic a rotation of the distant stars as seen by a freely falling -

hence locally inertial - observer. Thus in these models one might expect that the local kinematical inertial frame (determined by the fixed stars) and the local dynamical inertial frame would not agree. In that case they would represent models for what our Universe could have been like - or indeed might, in fact, be like within the limits of observational accuracy - but would not incorporate Mach's Principle.

Historically, anti-Machian solutions of the field equations were sought not by looking for solutions with shear, but by looking for solutions with rotation. The first such solution was found by Gödel (1949) in the case $\Lambda < 0$. The metric of Gödel's universe can be given in the form

$$ds^2 = -dt^2 + dx^2 - \tfrac{1}{2}exp(2\sqrt{2}\omega x)dy^2 + dz^2 - 2exp(\sqrt{2}\omega x)dtdy .$$

The constant ω (= $(4\pi\rho)^{\frac{1}{2}}$ = $(-\Lambda)^{\frac{1}{2}}$) is the magnitude of the relative rotation of material particles which flow along the t-coordinate lines. The universe is non-expanding and non-shearing.

Suppose an observer to be situated on one of the particles of the Gödel universe. Any particle may be chosen since the model has the same appearance to all such observers. One can ask with respect to what frame does this observer see nearby particles to be rotating? The answer turns out to be that the matter rotates with respect to a frame defined by the constancy of the direction of a local gyroscope; in other words, with respect to the frame in which Newtonian dynamics is locally approximately valid (without inertial forces). It follows that general relativity (with $\Lambda < 0$) admits cosmological models filled with matter in which the local kinematical and dynamical frames do not agree, and consequently cannot explain the observed coincidence of these frames as required by Mach's Principle (see also Gödel 1952).

This conclusion has been challenged on the grounds that the Gödel universe possesses some highly unphysical features more pathological than the violation of Mach's Principle. In particular there are closed time-like lines in the model, which rule it out as a physically reasonable model of the Universe. However entirely acceptable rotating models with no expansion are now known (Ostvath & Schücking, 1962). The difficulty in including expansion is that one must then introduce shear as well; this follows from a theorem due to Banerji (1968) and King (1972). No expanding, shearing, rotating models are known explicitly.

There are, however, tilted solutions known containing matter flowing with proper motion relative to those observers who see the geometry as spatially homogeneous. These models raise some interesting points. For one can find such models in which

160

true singularities exist which are not of the type we have discussed so far. In these tilted models there is an 'edge' to
space-time beyond which observers cannot travel (or from beyond
which observers cannot have come if the singularity is in the
past), at which the gravitational forces are quite finite. Thus
one cannot argue the irrelevance of the failure of general relativity in as much as it predicts the existence of space-time
singularities, where physics, hence general relativity itself,
breaks down, on the grounds that we know that before we get to a
singularity at which gravitational forces become strong enough
to affect the sub-atomic structure of matter, we must take
account of quantum physics to obtain a true theory of gravity.
For in these tilted models the classical theory should be valid
right up to the singularity! Such a boundary to space-time is
more serious than the limits to predictability we considered in
Section 9.3. Here our physics does not merely fail to tell us
how the world goes on; it appears to tell us that it simply
comes to a stop! (For different types of singularities, see:
Clarke & Schmidt 1977, Ellis & Schmidt 1972).

It is not a satisfactory solution of these problems to
simply discard those solutions we consider inappropriate. The
laws of physics themselves are supposed to tell us what cannot
happen; they should not require us to select the possibilities
we prefer for some external reasons, however 'physical' one might
think those reasons are. Thus if we postulate for example that
closed time-like lines are unphysical then no physical theory
should admit them. If it does, it fails to accord with the postulates and is to that extent a bad theory. At least with regard
to Mach's Principle, which is our main concern here, we should
like to rewrite general relativity such that only solutions satisfying some specified form of the principle are possible. We
shall come close to doing this, but in the end we shall be forced
to state the principle as a way of selecting solutions on a
systematic basis. Whether the systematisation of the selection
is adequate recompense for its *post hoc* nature is presumably a
question of aesthetics.

Chapter 10:

Mach's Principle and the Dynamics of Space-Time

The strongest principles in physics, and those most tenaciously adhered to, are principles of symmetry. It is difficult to contemplate a world which knows itself to be left- or right-handed, or has arbitrarily privileged directions and locations. But the manifestations of symmetry are not always obvious. Aristotle's view of the World is of a kinematic symmetry: the spherical heavens, organised around the radial symmetry of earthly things. Through Newton we find the motion of bodies accounted for by a dynamical law which is symmetrical with regard to orientation and location in space and time. For the law itself distinguishes no special directions or positions. Nor does it distinguish any special velocity, and in this way Newtonian relativity appears as a symmetry between observers in uniform relative motion.

It was just Einstein's reluctance to relinquish this symmetry of relative motion with respect to electrodynamic phenomena that led to special relativity. The relativity of general relativity can also be summarised in terms of symmetry, as the symmetry of gravity and acceleration and the consequent symmetry between all states of motion. But this symmetry of motion fails at the point where boundary conditions have to be introduced into the field equations of general relativity, and the theory fails to exhibit Mach's principle as the relativity of motion of matter with respect to matter.

At this level one can see the psychological power of Mach's principle: if symmetry forbids one to know that a sphere has been rotated, surely an observer on the surface of a sphere should not be able to discover that it is rotating. Thus, many attempts have been made to incorporate Mach's principle into a theory of gravity. To do this, one must either modify general relativity, or reinterpret Mach's principle, or both. In the following chapters we discuss three main types of approach.

If we describe Maxwell's equations as the dynamical theory of light, then we can summarise the development of our understanding of space-time structure succinctly as the appreciation of the changing arena for the action of dynamical laws. In general relativity this development is taken a stage further and the structure of space-time itself becomes the subject of dynamical theory. The Einstein field equations yield a geometry for space-time: from the four-dimensional space-time point of view, space-time is given at once, and not as an historical sequence. But, at least for stable causal solutions, the space-time may be assembled from spatial surfaces of constant cosmic time, in general in many ways. The geometry of these spatial surfaces changes from moment to moment in a way specified by Einstein's dynamical theory of geometry. From the point of view of this *3+1 decomposition* into space and time, there is a dynamical evolution. In Wheeler's (1962, 1964a,b) *geometrodynamics* the plan of this evolution, the precise way it is determined by the field equations, is regarded as the implementation of Mach's Principle in the theory.

The details of the way in which geometry evolves turn out to be very difficult to specify. This arises essentially from the fact that the metric coefficients contain information about both the geometry and the coordinates in which that geometry is described. The latter can be changed arbitrarily according to whim or insight; the former is fixed by the physics. Thus not all the metric coefficients can be determined by the field equations. Four of the field equations, a number that derives from the fact that there are four arbitrary functions describing the coordinates, act as constraints on the possible starting points for evolution. One can think of this as the memory of the Newtonian inverse square law acting instantaneously, and implying, from the start, the existence of certain forces appropriate to the matter distribution. There are parallels with Maxwell's equations in the theory of electromagnetism. General relativity needs four redundant functions to maintain explicit covariance, that is, manifest independence of choice of coordinates. Maxwell's theory

163

requires one extra potential to exhibit explicit covariance under Lorentz transformations. Maxwell's theory does not determine that potential, although it determines the physically measurable electric and magnetic fields correctly (Section 10.1). Einstein's equations do not determine the coordinate system, although they determine the physically measurable curvature correctly (Section 10.2).

Once a set of appropriate quantities can be specified, and their evolution traced, one can say that this represents the plan of general relativity. One can trace the evolution of the geometry of spatial surfaces in a *superspace* of three-dimensional geometries, and the plan provides for the construction of this superspace. In Wheeler's approach only closed three dimensional geometries are considered, in accordance with the suggestion of Einstein, since this eliminates the need to impose boundary conditions at spatial infinity. In this sense the theory is also a selection rule. It differs from other approaches (Chapter 12) in that, in a sense to be made precise by the theory, gravitational energy, as well as matter, is supposed to determine the geometry and hence the inertial properties of bodies.

The Brans-Dicke and Hoyle-Narlikar theories to be discussed in Chapter 11 are two variations on the theme of the possible dependence of the inertial mass of a body on the existence of other masses in the Universe. In the Brans-Dicke theory a variable strength of the gravitational interaction is introduced, and reinterpreted as a dependence of inertial mass on a long range interaction with other matter. Mach's principle appears as the requirement that the inertial mass should be wholly accounted for by this interaction. The Hoyle-Narlikar approach involves the reformulation of general relativity in terms of an action-at-a-distance theory in which the local physics is determined by the global properties of the Universe. Both theories fail to overcome the problems raised in Chapter 9.

In the third of these approaches we discuss the implementation of Einstein's form of Mach's principle as a rule for selecting acceptable solutions of the field equations. Of course, to be of any value, this has to be achieved in a systematic way. The required conditions can be related to an integral formulation of the field equations which explicitly expresses their global aspect.

We saw, however, that the apparent symmetry of the Aristotelian universe was an illusion, and disappeared on close inspection to be replaced by a more subtle underlying symmetry. Perhaps with regard to our present discussion too, the desire to impose a symmetry is frustrated because it should be imposed at a deeper level. In the final chapter we turn to some less developed ideas which take us beyond the frontiers of our understanding of space-time relativity.

§10.1 *The 'plan' of electromagnetic theory*

The theory of electromagnetism in Minkowski space-time, as expressed through Maxwell's equations, contains in a mathematically much simplified form, the essential problems to be faced in describing the dynamics of general relativity. We shall therefore consider electromagnetic theory in some detail and then (Section 10.2) develop the analogy with geometrodynamics without explicit computations.

The electric and magnetic fields are described by vectors $\underset{\sim}{E}$ and $\underset{\sim}{B}$ in the rest frame of some observer. The dynamical evolution of the electric and magnetic fields is described by Maxwell's equations in the standard form

$$\underset{\sim}{\nabla} \cdot \underset{\sim}{E} = 4\pi\rho \qquad\qquad \underset{\sim}{\nabla} \cdot \underset{\sim}{B} = 0$$

$$\underset{\sim}{\nabla} \wedge \underset{\sim}{E} = \frac{-\partial \underset{\sim}{B}}{\partial t} \qquad\qquad \underset{\sim}{\nabla} \wedge \underset{\sim}{B} = \frac{\partial \underset{\sim}{E}}{\partial t} + 4\pi \underset{\sim}{j} \quad ,$$

where ρ and $\underset{\sim}{j}$ are the charges and currents generating the fields. Explicit Lorentz covariance - the statement of the laws of electromagnetism in a form appropriate to any inertial observer in Minkowski space-time - is achieved by combining the electric and magnetic vectors into a single antisymmetric tensor, $F_{\mu\nu}$, according to

$$(F_{\mu\nu}) = \begin{pmatrix} 0 & E_1 & E_2 & E_3 \\ -E_1 & 0 & B_3 & B_2 \\ -E_2 & -B_3 & 0 & B_1 \\ -E_3 & -B_2 & -B_1 & 0 \end{pmatrix}$$

in the rest frame of the observer. Maxwell's equations then take the form

$$\partial_\mu F^{\mu\nu} = J^\nu \quad , \tag{10.1.1}$$

$$\partial_{[\lambda} F_{\mu\nu]} = 0 \quad , \tag{10.1.2}$$

165

where $\lfloor \lambda\mu\nu \rfloor$ indicates a sum of terms which is antisymmetric in any pair of indices - in this case equivalent to a sum over cyclic permutations. Explicitly, for example

$$\partial_{\lfloor 1} F_{23 \rfloor} \equiv \partial_1 F_{23} + \partial_2 F_{31} + \partial_3 F_{12} = 0 .$$

If we introduce a vector potential, A_μ, by

$$\partial_\mu A_\nu - \partial_\nu A_\mu = F_{\mu\nu} , \qquad (10.1.3)$$

equation (10.1.2) is automatically satisfied; in fact (10.1.2) is the necessary and sufficient condition for $F_{\mu\nu}$ to have the form (10.1.3). The remaining Maxwell equations (10.1.1), which relate the field to the currents J^ν become

$$\partial_\mu \partial^\mu A^\nu - \partial_\mu \partial^\nu A^\mu \equiv \Box A^\nu - \partial^\nu(\partial_\mu A^\mu) = J^\nu . \qquad (10.1.4)$$

The first term is just a definition of the d'Alembertian operator '$\Box$', and the second arises by changing the order of partial differentiation. Explicitly $\Box$ is just the wave operator $\left(- \dfrac{1}{c^2} \dfrac{\partial^2}{\partial t^2} + \dfrac{\partial^2}{\partial x^2} + \dfrac{\partial^2}{\partial y^2} + \dfrac{\partial^2}{\partial z^2} \right) .$

The solution of a problem in electrodynamics appears to reduce to finding the solution of (10.1.4), and substituting this in (10.1.3) to reconstruct the fields. However (10.1.4) does not determine a unique solution for given initial conditions. For if $A^\nu(x)$ is a solution, then so is $\overline{A}^\nu(x) = A^\nu(z) + \partial^\nu \Lambda$, where Λ is a completely arbitrary (differentiable) function. This follows by direct substitution. The transformation $A \to \overline{A}$ is called a *gauge* transformation. If we reconstruct $\overline{F}^{\mu\nu}$ from $\overline{A}^\nu$ we find $\overline{F}^{\mu\nu}$ = $F^{\mu\nu}$. So the arbitrariness in the potentials, the freedom to make gauge transformations, does not affect the measurable electric and magnetic fields. This does however mean that the 'plan' of electrodynamics cannot be described by saying 'solve (10.1.4)'.

To pass from what the plan is not to what it is, consider first what we should need to do if the equations had turned out to have the simpler form

$$\Box A^\nu = J^\nu , \qquad (10.1.5)$$

with J^ν unrestricted by a conservation law. In this case the arbitrariness in A^ν is restricted to the addition of a term $\partial^\nu \lambda$

where λ must now satisfy $\Box \lambda = 0$. This arbitrariness effectively disappears once we impose appropriate *initial conditions*, essentially because any solution of the 'wave equation', $\Box \lambda = 0$, which is initially zero must remain zero. We can therefore proceed to determine $A(t,x)$. The equation gives us $\frac{\partial}{\partial t}\left(\frac{\partial A^{\upsilon}}{\partial t}\right)$ in terms of $A^{\upsilon}(t,\underset{\sim}{x})$ and $J^{\upsilon}(t,\underset{\sim}{x})$. We assume that the currents, J^{υ} are given. Hence we obtain $\left(\frac{\partial}{\partial t}A^{\upsilon}\right)$ for all $\underset{\sim}{x}$, evaluated at $t +$ δt if we know A^{υ} and $\frac{\partial}{\partial t}A^{\upsilon}$ for all $\underset{\sim}{x}$ evaluated at t. We then obtain A^{υ} at $t + \delta t$ for all $\underset{\sim}{x}$, since $A^{\upsilon}(t+\delta t, \underset{\sim}{x}) = A^{\upsilon}(t,\underset{\sim}{x}) + \delta t \partial A^{\upsilon}$ $(t,\underset{\sim}{x})/\partial t$. To start the process we simply specify some initial values for $A^{\upsilon}(0,\underset{\sim}{x})$ and $\frac{\partial A^{\upsilon}}{\partial t}(0,\underset{\sim}{x})$ for all x on some initial surface $t = 0$ (say).

At each point, $\underset{\sim}{x}$, at $t = 0$ we specify four values of the dynamical 'coordinates' A^{υ}, and four 'velocities' $\partial A^{\upsilon}/\partial t$. This is like a dynamical system with four positions and velocities, so we say that the field has four degrees of freedom (at each point).

Now, in the case in which we actually have to solve (10.1.4), A^{υ} is undetermined up to a scalar function. We could eliminate this freedom by arbitrarily requiring one component of A^{υ} to have a specified form, say $A^1 = 0$, using the three independent equations contained in (10.1.4) to determine the three remaining unknowns. This is effectively what we do, but in a rather more 'planned' way: for aside from the inelegance of losing manifest Lorentz covariance, the singling out of a particular coordinate in the absence of any possibility of determining an explicit coordinate system would scarcely amount to a 'plan'. Instead, we use this freedom to put (10.1.4) in the form (10.1.5). For we can arbitrarily impose a single extra condition on A^{υ} to fix Λ. Suppose, for example, we impose

$$\partial_{\mu}A^{\mu} = 0 \quad , \tag{10.1.6}$$

the so called Lorentz condition. It is in fact sufficient to impose the Lorentz condition (10.1.6) only on the initial surface $t = 0$. For from (10.1.5), we derive a wave equation for $\partial_{\upsilon}A_{\upsilon}$:

$$\Box \partial_{\upsilon}A^{\upsilon} = \partial_{\upsilon}J^{\upsilon} = 0 \quad ,$$

the latter condition being the mathematical expression of the conservation of net electric charge. It follows from the nature of solutions of the wave equation that if $\partial_{\upsilon}A^{\upsilon} = 0$ at one time, it will vanish at all other times. If the initial A^v does not satisfy the Lorentz condition, we can find an $\overline{A}^v$ which does, by a gauge transformation determined by $\square\Lambda = -\partial_\mu A^\mu$. The problem is then posed in terms of $\overline{A}^v$ satisfying (10.1.4) and (10.1.6), which together yield (10.1.5).

However, we have not yet solved the problem because the constraint (10.1.6) is not the only restriction on the initial data. For if (10.1.6) is to be satisfied everywhere, by substitution for $\partial_0 A^0$ in the zero component of (10.1.5) we find

$$+ \ \partial_0 \partial_i A^i + \partial_i \partial^i A^0 = J^0 \quad .$$

This is, of course, precisely the zero component of the original system (10.1.4); since it contains no second time derivatives it constrains the initial data again! Note that if $\partial_0 A^i$ is regarded as known initial data, then this equation determines A^0 by an equation of Poissonian form - that is, by instantaneous propagation. This is nothing but the instantaneous $1/r^2$ electrostatic part of the electromagnetic field!

How can this instantaneous propagation be compatible with special relativity? Only if the electrostatic component has been present for all time, and if changes in the field are propagated with the speed of light. If we were to suddenly create a charge the electrostatic component would charge instantaneously and that would contradict special relativity. We must therefore be prevented from creating charges. This is expressed through the conservation of electric charge, $\partial_\mu J^\mu = 0$. From this relation it follows that the field equations are not all independent but are subject to a differential identity. One readily verifies that the left hand side of (10.1.4) has *identically* zero divergence, $\partial_v(\partial_\mu \partial^\mu A^v - \partial_\mu \partial_v A^v) \equiv 0$. But it is then precisely this conservation of charge which leads to the existence of gauge freedom, since the existence of a differential identity means that the equations are insufficient to determine the potential uniquely. We may expect that a theory of gravity which incorporates the $1/r^2$ force of Newtonian gravity in a relativistic form will be subject to precisely the same problem of gauge freedom which we have met here. In particular, general relativity, being a theory of the geometry of space-time, exhibits this gauge freedom in the arbitrariness allowed in the description of the metric, that is, in the freedom to make arbitrary coordinate transformations.

We have learnt that our attempt to reduce the system (10.1.4) to the determinate form (10.1.5) imposes constraints on the data; these we must now show how to solve. For convenience, we give up manifest Lorentz covariance and work with a somewhat restricted coordinate system adapted to the surfaces t = constant on which the dynamical evolution of the electromagnetic field is to be presented. Thus, we separate the four-potential, A^v, into a three-vector, $\underset{\sim}{A}$, and a scalar, $\phi \equiv A^0$ on the spatial surfaces.

The Lorentz gauge condition (10.1.6) involves not the whole of A^v but only ϕ and the part of $\underset{\sim}{A}$ with non-zero divergence. Consequently we split $\underset{\sim}{A}$ into a transverse component, $\underset{\sim}{A}^t$, defined by the requirement that its divergence vanish, and a longitudinal part, $\underset{\sim}{A}^\ell$, with non-zero divergence. This is possible, since for $\underset{\sim}{A}^t$ we can take the curl of an arbitrary vector, $\underset{\sim}{A}^t = \underset{\sim}{\nabla} \wedge \underset{\sim}{W}$, and for $\underset{\sim}{A}^\ell$ we can take the gradient of a scalar, $\underset{\sim}{A}^\ell = \underset{\sim}{\nabla} \psi$. The vector $\underset{\sim}{A}^t$ contains two pieces of information (since it is constrained by $\underset{\sim}{\nabla} \cdot \underset{\sim}{A}^t = 0$), and $\underset{\sim}{A}^\ell$ contains one (viz. ψ).

Note that $\underset{\sim}{A}^t$, the transverse part of $\underset{\sim}{A}$, is unaltered by a gauge transformation, which is accommodated entirely by changing ψ to $\psi + \Lambda$. The transverse potential is therefore a candidate for representing a physically measurable quantity: in fact, it is essentially equivalent to the magnetic field, $\underset{\sim}{B}$. This follows since $\underset{\sim}{B} = \underset{\sim}{\nabla} \wedge \underset{\sim}{A} = \underset{\sim}{\nabla} \wedge \underset{\sim}{A}^t$. Thus, the initial conditions for dynamical evolution of an electromagnetic system involve the specification of $\underset{\sim}{B}$ subject to the constraint that $\underset{\sim}{\nabla} \cdot \underset{\sim}{B} = 0$: and this is equivalent to the free specification of the two arbitrary functions in $\underset{\sim}{A}^t(x,0)$.

For the electric field, we have $\underset{\sim}{E} = -\dot{\underset{\sim}{A}} - \underset{\sim}{\nabla}\phi$; hence $\underset{\sim}{E}^t = -\dot{\underset{\sim}{A}}^t$, and the initial specification of the velocities, $\dot{\underset{\sim}{A}}^t$, which involve two arbitrary functions, is provided by the two pieces of information contained in the divergence-free part of the initial electric field. The longitudinal part of $\underset{\sim}{E}$ must satisfy $\underset{\sim}{\nabla} \cdot \underset{\sim}{E}^\ell = 4\pi\rho$, so is essentially completely determined by the charges (and the boundary conditions at spatial infinity). Thus, four arbitrary functions are involved in the initial specification of the physical state of the system, two to describe the 'coordinates', $\underset{\sim}{A}^t(0,\underset{\sim}{x})$, at each point, and two for the velocities, $\dot{\underset{\sim}{A}}^t(0,\underset{\sim}{x})$. This is described by saying that the electromagnetic field has two degrees of freedom.

From the dynamical equations (10.1.4) the evolution of the transverse potential is determined. We find

$$\Box \underset{\sim}{A}^t = \underset{\sim}{J} - \Box \underset{\sim}{A}^\ell \ . \qquad (10.1.7)$$

To determine $A^\ell_\sim$, we use the Lorentz conditions (10.1.6) in the form $\nabla^2 \psi + \dot{\phi} = 0$, and the remaining evolution equation, the zero component of (10.1.4), which now reads

$$\nabla^2 (\dot{\psi} + \phi) = \rho \quad .$$ (10.1.8)

Thus, eliminating ϕ gives $\nabla^2 (\Box \psi) = \dot{\rho}$, or, symbolically, $\Box \psi = 1/\nabla^2 \dot{\rho}$, from which we can reconstruct $\Box A^\ell_\sim$ as $\Box A^\ell_\sim = \Box \nabla \psi = \nabla \Box \psi = \nabla (1/\nabla^2) \dot{\rho}$. Consequently the equations for the dynamical degrees of freedom become simply

$$\Box A^t_\sim = J_\sim - \nabla (1/\nabla^2 \dot{\rho}) \quad ,$$

and there is no need to determine $A^\ell_\sim$ explicitly. Equation (10.1.8) now gives $\dot{\psi} + \phi = 1/\nabla^2 \rho$, from which we can reconstruct the longitudinal electric field immediately, since $E^\ell_\sim = \nabla (\dot{\psi}+\phi) = \nabla (1/\nabla^2 \rho)$.

This therefore completes our specification of the 'plan' of electrodynamics. We have specified what data is to be given freely ($B_\sim$, $E^t_\sim$ initially, and the distribution and flow of charge) and what is thereby determined (the evolution in time of the electromagnetic field). We should note that we have here ignored the fact that in practice we do not know the currents and charge densities until we have solved for the fields, since the fields determine the motion of particles and hence the charges and currents. This does not alter the logic of the 'plan' but considerably complicates its practical implementation.

The plan of electrodynamics is, in fact, far from unique. One might guess that since $A^\ell_\sim$ has played such a minor role in the above discussion it could be eliminated altogether. This is indeed often the method used. In place of the Lorentz condition we impose the *Coulomb* gauge condition, $A^\ell_\sim = 0$. Then ϕ is determined by $\nabla^2 \phi = \rho$, or $\phi = 1/\nabla^2 \rho$ everywhere, and the transverse potential is given by $\Box A^t_\sim = J_\sim + \nabla \dot{\phi}$. A further alternative approach is possible in which no gauge condition is imposed, but instead ϕ is chosen arbitrarily for all time. The zero component of the equations of motion (10.1.4) then play the role of a gauge condition which can be solved for ψ, and the complete solution then determined by specifying the free data in $A^t_\sim$.

The extensive freedom of choice here is related to the linearity of the dynamical equations, which are therefore readily seen to possess acceptable solutions. In general relativity there is a high degree of nonlinearity in the field equations so that all choices of gauge may not be equally possible, and only for suitably cunning choices have the resulting systems been shown to be soluble.

§10.2 *The 'Plan' of General Relativity*

The metric contains information about the coordinates because it gives the measured distances between points with different coordinate values, the coordinate values themselves being arbitrary numbers. There are many ways in which this information can be extracted from the metric. For example, one could require the coordinates to be specified physically, with time being chosen as the proper time of a class of observers in non-rotating motion. Choosing the class to follow trajectories orthogonal to the spatial sections, the geometries of which are to represent the dynamical states of the system, we specify the Gaussian coordinate system with $g_{oo} = -1$, $g_{oi} = 0$. However, such coordinate systems tend to break down, becoming singular a finite time into the future of the initial surface. Consequently a better approach is to isolate the coordinate information in the metric coefficients, but to keep this unrestricted.

It turns out to be appropriate to choose just the metric coefficients g_{oo} and g_{oi} as the coordinate labels for the development of the coordinate system off of the initial surface. This choice is dictated by the nature of the field equations, just as the choice of $\phi = A^o$ as a non-dynamical quantity in electrodynamics is imposed essentially on the basis of the fact that its second time derivative never appears in the electrodynamic field equations. In principle one can verify, by writing out the Einstein equations explicitly, that the second time derivatives of g_{oi} and g_{oo} do not appear. In practice the analysis is carried out more simply in terms of a Lagrangian for the metric field, since this is a single scalar quantity giving rise to the ten field equations. The Lagrangian differs from those of Newtonian dynamics (Section 3.6) in that the dynamical quantities are labelled by a continuous variable, $\underset{\sim}{x}$, and are therefore infinite in number, in contrast to the finite number of coordinates required to determine the dynamical evolution of a finite number of particles. But this is more in the nature of a technical difference than one of principle.

Since four functions of coordinates are arbitrary it follows that four of Einstein's equations serve to constrain the initial data rather than to determine its evolution. This again means that the system of field equations is indeterminate, and coordinate conditions must be specified in order to obtain a solution. The problem arises here also from the existence of conservation laws, the conservation of energy and momentum in this case, and the consequent existence of differential identities amongst the field equations. These identities guarantee that the constraints once satisfied remain satisfied, as is appropriate to their origin as conservation laws. There is, however, a significant difference from electrodynamics. The creation of net charge is absolutely forbidden; the creation of mass or energy in the form of matter is permissible, since the energy can be extracted from the gravitational field. Electric charge cannot be extracted from the electromagnetic field, since a light beam carries no charge; on the other hand, a gravitational wave carries mass and energy, and we have already noted that it is this that leads to the non-linearity of Einstein's equations in contrast to the linearity of Maxwell's system.

The indeterminacy of the field equations, and the apparent violations of causality associated with the ghost of the inverse square law, at first led Einstein to suspect the principle of covariance to be at fault, and to seek non-covariant equations. It was Hilbert (1915) who apparently first gave the correct analysis of the situation. From our point of view we see the need to specify a coordinate condition, analogous to the gauge condition of electrodynamics, in order to spell out the dynamical evolution of a metric 3-geometry.

Following Arnowitt, Deser and Misner (1962), it is conventional to choose the notation $g_{oi} = N_i$, $N^i = g^{ik}N_k$, $g_{oo} = N_iN^i - N^2$, so

$$ds^2 = g_{\mu\nu}dx^\mu dx^\nu = -(Ndx^o)^2 + g_{ij}(N^i dx^o + dx^i)(N^j dx^o + dx^j) \ .$$

The significance of this choice is expressed by noting that Ndx^o represents the proper time between the surfaces x^o = constant and $x^o + dx^o$ = constant, and $N^k dx^o$ represents the displacement in the upper surface of a point with the same spatial coordinate labels on both surfaces (Fig. 10.1).

The components g_{ij} represent the geometry of the spatial sections and therefore contain the dynamical quantities, the time evolution of which is governed by the field equations. Of course, the coordinates in the spatial surfaces are also arbitrary, and this means that not all the components of g_{ij} represent true dynamical quantities. The rate of change of g_{ij} with time, $\partial g_{ij}/\partial x^o$, is related to the change in time of the 3-geometry of the spatial sections. It cannot be equated with the changing geometry since a fixed geometry (ordinary Euclidean spaces, for example) expressed in time dependent coordinates will give rise to a non-zero $\partial g_{ij}/\partial x^o$. In a coordinate independent way, the changes in geometry from one surface to the next are represented

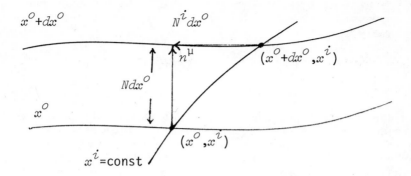

Fig. 10.1. $N^i dx^0$ gives the point where the normal n^μ to the surface labelled by x^0 pierces the upper surface labelled by $x^0 + dx^0$; $N dx^0$ gives the length of the normal.

by the convergence or divergence of the normals $n_\mu = (N,0,0,0)$. The vector obtained by parallel transport in space-time from P to a neighbouring point Q in a given surface can be compared with the normal at Q, and the difference used as a measure of the convergence or divergence (Fig. 10.2). This defines the ex-

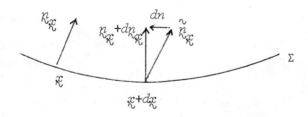

Fig. 10.2. The vector $dn_i = -K_{ij}dx^j$ defines the way the surface Σ is embedded in a space of one higher dimension. n_x is normal to Σ at x, and $\tilde{n}_x$ is obtained from n_x

by parallel transport in the embedding space.

173

trinsic curvature K_{ij} of a surface, the way the 3-surface is embedded in space-time, by $\delta n_i = -K_{ij}\delta x^j$. Explicit computation yields an expression for K_{ij} which depends on N, N_i as well as $\partial g_{ij}/\partial x^o$, namely

$$K_{ij} = \tfrac{1}{2}N(\partial N_i/\partial x^j + \partial N_j/\partial x^i - \partial g_{ij}/\partial x^o) + N\Gamma^k_{ij}N_k$$

$$= K_{ji}$$

One can think of the six independent components of K_{ij} as the geometrical analogue of the three electric field components E_i, while the six g_{ij}'s take the role of the three components of the vector potential A_i. We must now isolate from g_{ij} and K_{ij} the dynamical variables which are to represent the 3-geometry of the spatial sections. One might, for example, insist that the coordinates be chosen to make g_{ij} diagonal, thereby using up all the freedom to choose 3-space coordinates. This turns out to be inappropriate for the development of a general theory of the phase-space dynamics of general relativity, but this approach may be borne in mind as adequate to provide an understanding of the nature of the argument. There is a further coordinate freedom in the choice of time, since not only must the intrinsic geometry of a 3-space be determined, but also its location in space-time. In this sense, the three-geometry is the carrier of information about time, information which is contained in K_{ij} (Wheeler, 1967). There are different approaches to extracting this information.

Three of the four constraints restrict K_{ij} essentially by specifying its divergence:

$$(K^j_{\ i} - \delta^j_{\ i}K^m_{\ m})|_j = \kappa T^o_{\ i} \tag{10.2.1}$$

where $|_j$ denotes the covariant derivative in the initial 3-surface defined by parallel transport in that surface. This leaves three pieces of information in K_{ij}. Together with the three remaining metric coefficients, we have six variables to specify. But still to be subtracted is the one remaining constraint and the freedom in the choice of time. This leaves four pieces of information (functions of position at the initial time) to be specified freely; two of these are dynamical 'coordinates' (parts of g_{ij}), and two of them are the corresponding initial rates of change, or 'velocities' contained in K_{ij}. The future development of the dynamical variables is controlled by the field equations. This is analogous to the four arbitrary functions in the initial values of $\underset{\sim}{A}^t$ and $\underset{\sim}{\dot A}^t$. Thus the electrodynamic field and the gravitational field both have two true degrees of freedom.

In Wheeler's approach one does not explicitly extract the information about time. The plan of general relativity is described in the following way. Pick arbitrarily g_{ij} and $\partial g_{ij}/\partial x^0$ on the initial surface. This fixes the choice of coordinates and their initial development in time. The four constraints may then be regarded as initial value equations for N and N_i. Thus we have all the initial values we need. The dynamical equations do not determine N, N_i off the initial surface, so these must now be specified. Once N, N_i are chosen, the Einstein equations determine uniquely the evolution of g_{ij} and K_{ij}, hence the geometry of space-time.

There are some difficulties associated with this approach. Essentially g_{ij} and $\partial g_{ij}/\partial x^0$ on the initial surface are not entirely arbitrary, but must contain an appropriate definition of 'time' specifying the location of the hypersurface compatible with its embedding. Unless this happens the constraint equations will not be solvable for N and N_i. That consistent data can be given such that the constraints can be solved uniquely for the initial values of N and N_i is known as the 'thin-sandwich conjecture'. Partly because of the non-standard form of the initial value equations for N and N_i no decisive results have been obtained. (But see Fischer and Marsden, 1979, p.148.)

If one accepts the conjecture then one can state a version of Mach's Principle in general relativity:

> the specification of a 3-geometry and its time
> rate of change on a closed space-like surface
> together with the energy density and density of
> energy flow on the surface determines the geometry
> of space-time and hence the inertial properties of
> test particles.   (Wheeler, 1964c)

Implicit in all of this, of course, is the assumption that the space-time geometry obtained is globally determined by the data on an initial hypersurface. This restricts us to space-times which are globally hyperbolic as discussed in Section 12.3. In general this 'plan' will yield only a part of space-time.

This approach may be looked at from another point of view through the construction of the configuration space appropriate to the dynamics of space-time geometry. This is Wheeler's *superspace* which consists of the set of all positive definite three-geometries on closed surfaces, each of which is a point in superspace (Wheeler, 1968). Each three-gometry is specified by a metric, together with all the different forms that metric can take when expressed in different coordinate systems. Therefore each three-geometry is defined by three pieces of information per space point (six metric components less three arbitrary co-ordinate choices). The restriction to positive definite metrics, by which we mean metrics that can be reduced by a coordinate transformation to the Euclidean form at any one point, is ob-viously necessary in the classical theory, since we require space to be locally Euclidean. This requirement might have to be relaxed in a quantum theory; allowing metrics of arbitrary signature one obtains *extended superspace*. The restriction to closed (compact) surfaces is inserted mainly for mathematical convenience, since it obviates the necessity to specify boundary conditions at spatial infinity, and simplifies the proofs of certain theorems. The evolution of a closed three-geometry according to Einstein's equations is represented by a curve in superspace.

Superspace is not a differentiable manifold but is very like one (Stern, 1967). For it turns out that spaces with spec-ial symmetries have neighbourhoods which cannot be continuously deformed to look like neighbourhoods of metrics lacking any symmetry. The 3-manifolds with special symmetry are therefore singled out, and lie on the boundaries of the set of manifolds without symmetry. Such a structure is called a *stratified mani-fold* (Fisher, 1970). However, the relevant point here is that superspace possesses sufficient differentiable structure that we can define tangent vectors to curves in superspace. In particular, given a metric for superspace, we can define geodesics. DeWitt (1970) has shown that superspace can be given a metric in which the curves of solutions of Einstein's equations are geodesics.

These considerations rest at least partly on the unproved validity of the thin sandwich conjecture. It is now possible to avoid this element of uncertainty by an alternative approach due to York (1971, 1972, 1973) and based on work of Lichnerowicz (1955) and Bruhat (1962) (see also Fischer and Marsden, 1979). In York's approach one makes a particular choice of the variable which is to represent time. The variable chosen is essentially the trace $(K^i{}_i)$ of the extrinsic curvature. This trace repres-ents the rate of change of volume per unit volume per unit time, hence the rate of expansion of the geometry at a point. In this sense it generalises the Hubble constant. If we call the app-ropriate trace τ, the initial hypersurface is then labelled $\tau =$ constant. This leaves two free dynamical components of K_{ij} when one takes account of (10.1.8). Furthermore an explicit decomp-osition of K_{ij} into free and constrained parts can be given, as was the case for $\underset{\sim}{A} = \underset{\sim}{A}^t + \underset{\sim}{A}^\ell$ (see also O'Murchadha and York,

1974).

The remaining constraint is required to determine the conformal scale of the three-gometry; one specifies freely the three-gometry up to an overall multiplicative function. This clearly requires two pieces of information - the ratios of the metric coefficents in the case where the metric is chosen to be diagonal. This leads to a constraint of standard elliptic form which can be shown to possess a unique solution for closed spatial geometries, or for appropriate boundary conditions at infinity. This identification of the conformal factor as the non-dynamical variable was first made by Dirac (1959, 1964). However, he chose a time variable by demanding that the 3-spaces be embedded in space-time in such a way that $\tau = 0$ for all time, and such a choice is now known to be impossible in a closed universe.

In view of the fact that the three-gometry determined by g_{ij}, and K_{ij} carries information about time, Wheeler was led to ask, 'What is two-thirds of superspace?' The answer given by York is 'conformal superspace', the collection of three-dimensional Riemannian metries defined up to a conformal factor. York has shown how a physically measurable curvature may be constructed from a conformal geometry in a way that essentially parallels the construction of the physically measurable magnetic field from the transverse part, $A^t_{\sim}$, of the vector potential.

§10.3 *Mach's Principle and Geometrodynamics*

The relation of the above analysis to Mach's Principle has been spelled out by Isenberg (1974). First, Isenberg claims that the explicit presentation of the plan of general relativity provided by York implements Wheeler's form of Mach's Principle (apart, possibly, from certain presumably purely technical difficulties in the case $T^0_{\sim i} \neq 0$). Second, it provides a method for the explicit construction of model universes satisfying Mach's principle. These models include both empty space-times and homogeneous cosmologies with rotation! With regard to the former it is argued that the space-times are not empty since they contain gravitational radiation. This interpretation does rule out the genuinely empty Minkowski space-time since this is not spatial closed, and it rules out the flat, empty, spatially closed torus obtained by 'rolling up and glueing' Minkowski space-time, since the scale of this is shown to be undetermined in York's analysis. In the case of homogeneous rotating cosmologies, local rotation is claimed not to be an anti-Machian property, since the local inertial frame is indeed determined by the matter and energy -

177

including gravitational energy - in the Universe.

The justification for including as satisfying Mach's Principle models which are often considered to be anti-Machian rests, therefore, entirely on the contention that it is appropriate to consider gravitational energy as a source of inertial forces. One can consider this directly in the case of Wheeler's gravitational 'geons' (Wheeler, 1955, 1962a). These were originally conceived as pure electromagnetic energy (or neutrino energy) held together by its own gravitational attraction so as to be stable, or at least long-lived entities. The idea has been extended (Hartle, 1960; see also Fletcher, 1962) to purely gravitational geons which consist entirely of gravitational radiation. We shall consider only these gravitational geons. The radiation must have wavelengths much less than the size of the geon in order to be contained, and hence much less than the length scale over which the space-time curvature changes significantly, since this will be of the order of the geon size. Isaacson (1968) has shown how to separate, in a physically meaningful manner, a high frequency, short wavelength component of gravitational radiation from a slowly varying background, in order that the radiation may stand explicitly as the source of the large scale curvature, on the right hand side of the Einstein equations. From this point of view the geon appears on exactly the same footing as a material source of gravity, despite the fact that it consists of empty curved space-time. A distant observer could not tell the difference between such a gravitational geon and a material source, and since Mach's principle is usually taken to refer to distant sources, it can be argued that it follows that gravitational energy must be included as a source. However, Mach's principle in the form alluded to states that the inertial properties of every and any test particle must be determined by gravitating 'matter' in the Universe. In the present case the inertial properties of a particle inside the geon will not be so determined, since the gravitational energy therein is not a source of the geometry but is itself the geometry. For such a particle the space-time is manifestly devoid of sources of gravity in the sense of Mach's principle. This argument for including gravitational energy as a source is therefore by no means conclusive.

One can point also to the lack of sensible distinction between a field and its source in, for example, the early Universe. As we go back in time we see the Universe winding up again. Distances get smaller between clusters of galaxies, then between galaxies and, initially, before the formation of stars and galaxies, between the atoms themselves. The system heats up as

electromagnetic radiation is compressed to smaller volumes,
until a point is reached at which the radiation is in equilibrium
with matter, allowing a continual interchange of energy between
particle-antiparticle pairs and radiation. A particle and an
antiparticle can be created together from radiation energy, and
annihilated into radiation on collision. In this situation one
cannot say that the matter is the source of the electromagnetic
field or the field the source of the matter. One can say only
that one has a state of equilibrium with the rates of conversion
between the two forms being equal. We may assume that earlier
in time a similar situation prevailed with the gravitational
field - an equilibrium between gravitational and material and
radiation energy. From this point of view we cannot distinguish
between matter as the source of gravitational energy and gravitational radiation as the source of matter. The state of
excitation of the freely specifiable gravitational degrees of
freedom must simply be fed in as an initial condition, and may
be taken to be non-zero. Indeed, if Mach's principle were to
require a zero level of excitation of the gravitational modes,
one could ask whether this requirement should not then apply
equally to all forms of radiation (electromagnetic, various types
of neutrino) and to all types of massive particles. In that case
there would be no Universe, and this would at least be an observationally refutable consequence of Mach's Principle.

The argument depends on an analogy between gravity and
other forms of energy. This is certainly not justified at the
classical level, where the gravitational force appears to be
different from all other forces, which are not geometrical in
origin. Since we do not yet possess a quantum theory of gravity
it is impossible to decide whether the analogy would be possible
at that level. Thus, as far as we know, it is perfectly permissible to maintain a distinction between the *gravitational* field
and its sources. Once this distinction is admitted, Wheeler's
program is seen to be incomplete. One must add some conditions
on the level of excitation of the gravitational wave modes in the
Universe which would relate them to energy emitted by *material*
sources in the past.

Finally we should mention that the closure postulate
appears to be in some danger from observations. Suppose it were
to be shown that the Universe is indeed not closed. If the closure postulate is essential to the plan of general relativity,
then the observations would show general relativity to be wrong.
If the closure postulate is not essential - and this would be a
consistent way out of the dilemma - then Mach's principle as the
plan of general relativity admits open models. Either these open

models include asymptotically flat space-times, which are
patently anti-Machian, or some further selection rule must be
imposed. In the latter case, as far as Mach's Principle is concerned, the programme is clearly incomplete. Nevertheless, it
should be clear that the elucidation of the 'plan' of relativity
as the dynamics of spatial geometry has played a vital role,
particularly in the development of a quantised version of general
relativity, and is still the subject of research that is unfold-
ing for us the intrincate structure of general relativity. To
appropriate the name 'Mach's Principle' for this enterprise is a
serious abuse only if it is intended to be exclusive.

Finally, we might note that Wheeler has pursued an independent line of development of these ideas in which it is proposed
that the Universe consist entirely of space-time geometry. The
essence of this programme is to regard Einstein's field equations
as providing an 'already unified theory' of classical physics in
the sense that the apparently material fields which appear as
sources in the equations have a characteristic imprint on the
space-time geometry from which they can be reconstructed. At
the basis of this idea is the fact that Einstein's equations
alone provide not only the equations of evolution of the geometry,
but also the equations of evolution of matter, without further
assumptions (Section 8.6). In this way it is proposed to reduce
the motion of matter to the dynamics of geometry. Electrically
charged particles would have a different internal geometry from
neutral ones and hence a different motion. There are difficulties if other fields are included, neutrinos for example, for it
appears that these cannot be uniquely reconstructed from the
geometry. One might regard this form of physical monism as the
antithesis of Machian ideas, in that a logically privileged
position is assigned to empty space-time rather than to matter:
space-time generates matter rather than the matter generating
space-time.

Chapter 11: Theories of Inertial Mass

In general relativity the mass of a body is a property of the body and exists independently of any other bodies in the Universe. Mach's Principle then requires that the accelerations, equivalent to inertial forces, should be measured relative to the bulk of matter in the Universe. There is, however, an alternative possibility which can be stated very simply in an argument due to Narlikar. Suppose the Universe to contain a single isolated body subjected to no forces. Newton's law, $ma = F = 0$, appears to require the body to possess no acceleration in some absolute sense. Kinematically however the acceleration is quite arbitrary since it may be measured in an arbitrary reference frame. Narlikar observes that an arbitrary acceleration is compatible with the dynamics if $m = 0$. Consequently, in order to implement Mach's Principle, one is led to look for a theory in which inertial mass itself is generated by an interaction with other bodies in the Universe in such a way that the mass of a single isolated body is zero. The theories of Brans and Dicke (1961; see also Brans, 1962) and of Hoyle and Narlikar (1974), which we discuss in this chapter, represent attempts to implement this point of view.

It might appear that a theory which relates the mass of a body to the distribution of matter in the Universe should allow for a directional dependence of a locally measured mass associated with the anisotropic distribution of matter in the Galaxy. Cocconi and Salpeter (1958, 1960; see also Carelli, 1959) suggested that the inertial mass may be considered as the sum of an isotropic part, m, and an anisotropic part, Δm, and that the contribution to the mass of a body due to a mass M at distance r is proportional to Mr^{-v} with $0 < v < 1$. If local anisotropy is to be attributed to the Galaxy, mass M_G, radius R_G, and the isotropic part to the remaining matter in the visible Universe, density ρ out to radius R, we have

$$\frac{\Delta m}{m} = \frac{M_G}{R_G^{\,v}} \frac{3-v}{4\pi\rho R^{3-v}} \quad .$$

This yields $\Delta m/m$ ranging from 2×10^{-5} for $v = 1$ to 3×10^{-10} for $v = 0$.

Experiments to test this prediction were carried out by Hughes et al. (1960), Beltran-Lopez et al. (1961) and Sherwin et al. (1960) and with even greater accuracy by Drever (1961), (for a review see Hughes, 1964). A null result was obtained.

Dicke (1964, pp. 14-22) has argued that this is not to be interpreted as ruling out inertial mass theories of Mach's Principle, but as evidence that the anisotropy of inertial mass in such theories must be universal, the same for all particles. Dicke (1961) points out that if inertial mass has the form $m_{\mu\nu} = mf_{\mu\nu}$ with $f_{\mu\nu}$ a field generated by distant matter and independent of bodies responding to it, then the kinetic energy of a particle has the form

$$\tfrac{1}{2}\, mf_{\mu\nu}\, \frac{dx^\mu}{ds} \frac{dx^\nu}{ds} \quad .$$

Hence, $f_{\mu\nu}$ plays the role of a metric and consistent equations of motion can be obtained in which particle paths are geodesics in the metric $f_{\mu\nu}$. It follows that, as in general relativity, one may transform to a 'freely falling' frame in which $f_{\mu\nu}$ is just diag $(-1,1,1,1)$, and the paths of particles are independent of $f_{\mu\nu}$. While this argument is correct as far as it goes, it is not clear why $f_{\mu\nu}$ should play the role of a metric for electrodynamic or nuclear phenomena also, and consequently why anisotropies should not show up in the behaviour of light or in the action of nuclear forces.

Indeed, Dicke appears to argue just this point in using the null result of the Hughes-Drever experiment to set limits on the affect of any spatially anisotropic matter. Suppose in a freely falling frame certain directions were to be picked out. These might be described by a second rank tensor $h_{\mu\nu}$, since only by a special choice of directions of basis vectors can a general $h_{\mu\nu}$ be put into diagonal form. If matter is to couple to the field $h_{\mu\nu}$ there must be some parameter, λ/m say, which measures the strength of this coupling to a body of mass m, that is, which enables one to construct a force from the value of the field. The null result of Drever's experiment enables one to state that the ratio of λ/m to the strength of coupling of the gravitational field, measured by $(GM/c^2R)^{\frac{1}{2}}$, is less than 4.5×10^{-23}, and hence entirely negligible.

This important result rules out the possibility of a long range interaction for the generation of inertial mass mediated by a second rank tensor field. On the assumption that no higher rank tensor will be involved, this leaves either a vector or scalar field. Difficulties appear if one tries to introduce a vector field to represent a purely attractive force by changing the sign of the electromagnetic interaction, since the resulting theory has no lower bound to the energy, hence no ground state. The simplest generalisation is therefore the introduction into general relativity of an additional long range scalar field.

§11.2 *The Brans-Dicke theory*

The way in which the additional scalar field enters the Brans-Dicke theory has its origins in certain cosmological considerations of Dirac (1938; see also 1973a,b). Out of quantities which play an important role in physics such as, for example, the mass, m, of the proton, the strength of gravity as measured by the gravitational constant G; the value of the Hubble 'constant' H, and Planck's constant, h, from quantum theory, one can construct certain dimensionless ratios. For example, $m(G/hc)^{\frac{1}{2}}$ where c is the speed of light, is dimensionless, hence independent of the units in which the constants are expressed, provided these are chosen consistently. This dimensionless number expresses the ratio of the mass of a proton to the so-called Planck mass, $(hc/G)^{\frac{1}{2}}$, which may play a role in the quantum theory of gravity. A further example is the ratio (mc^2/hH), which is approximately the age of the Universe expressed in units of the time it takes light to travel across a proton (the size of which is taken to be

183

its *Compton wavelength*). Now Dirac argued that a complete physical theory should predict the magnitudes of these and other similar numbers, not merely take them as given data. On the other hand he noted that these number, or their reciprocals, are all very large, of order 10^{20} and 10^{40} in the case of the two quoted, and it is unreasonable to expect to be able to construct such values out of numbers which occur naturally in physics, such as π and e. One might therefore expect that these large numbers are related in some way, so that they do not have to be determined independently, and indeed one finds that to an order of magnitude approximation simple relations exist. We have, for example, with $H \approx 100$ *km/sec/Mpc* $\approx 3 \times 10^{-18}$/*sec*,

$$\left(\frac{hc}{G}\right)^{\frac{1}{2}} \frac{1}{m} \approx \left(\frac{mc^2}{\hbar H}\right)^{\frac{1}{2}} \approx 10^{20} \quad .$$

The second member of this relation depends on time through Hubble's 'constant', so if this relation is to remain true for all epochs, and not merely appear as an accident of the stage in its history at which we view the World, the first member must also depend on time. This led Dirac to suggest a cosmology in which the gravitational constant G is replaced by a function of time to be determined in the theory.

The theory was taken up and developed in modified form by Jordan (1955, 1959, 1973) whose results bear some relation to the Brans-Dicke theory. However, Jordan dealt with a field variable which is reciprocal to that to be introduced in the Brans-Dicke theory, and limited the material content of the Universe to the electromagnetic field only. Furthermore the continuous creation of matter adopted by Jordan is dropped in the Brans-Dicke version.

In order to decide upon an equation to determine the variation of G, we return to the relation

$$\frac{8\pi}{3} \frac{GM}{Rc^2} = 1 \quad ,$$

which we found for the Einstein-de-Sitter model universe in Chapter 9. This relation holds exactly only in that particular model, but is valid to order of magnitude for model universes which do not differ too much from the one which we appear to inhabit. We can consider this as an expression for the determination of G by the mass and radius of the visible - hence causally related - Universe. If G is to be so determined then dimensional arguments lead directly to a relation of this form. We can think

184

of it as a sum over the distribution of matter determining G^{-1}, namely as

$$G^{-1} = \frac{1}{c^2} \sum \frac{m}{r} \quad ,$$

where m is the mass and r the distance of a particle, and the sum is over all particles in the visible Universe. This suggests, with suitable relativistic generalisation, that $\phi \equiv G^{-1}$ should satisfy a linear wave equation.

To obtain a consistent set of equations, Brans and Dicke start from an action principle, which they choose to be of the form

$$\delta \int \left(\phi R + \frac{16\pi}{c^4} L_m - \omega \phi,_\mu \frac{\phi^{,\mu}}{\phi} \right) (-g)^{\frac{1}{2}} d^4 x = 0 \quad .$$

Here L_m is the Lagrangian density describing the matter and ω is a dimensionless constant, which measures the importance of departures from general relativity. In the limit $\omega \to \infty$ one regains Einstein's theory.

The ϕ field appears in the field equations in place of the gravitational constant, and also contributes to the stress-energy determining the metric, hence as an additional term on the right hand side of the field equations. The ϕ-field is generated by the trace of the stress-energy tensor of the matter fields through

$$\Box \phi = \frac{8\pi}{c^4} (3 + 2\omega)^{-1} T \quad . \tag{11.2.1}$$

To relate this to our initial considerations regarding the global determination of inertial mass, consider what it means to state that the gravitational 'constant' is changing. In order to measure G we have to set up units of mass, length and time. No physical quantity can be measured absolutely in any sense, but must be related to some arbitrary standard. Thus we measure not the mass of a body exactly, but its ratio to the standard kilogram, or to the mass of the electron, or to the Planck mass, etc. Similarly with other quantities. One cannot continue indefinitely however, because the values of certain dimensionless ratios are fixed for us by Nature, quite independently of the choice of units. We have already seen that $(G/\hbar c)^{\frac{1}{2}} m$ is a fixed number, so once we have fixed the units of three of these quantities appearing in this expression, the fourth is determined. Clearly, therefore, it

185

is strictly meaningless to state only that the gravitational constant is changing without simultaneously specifying the choice of units, for we are free to choose units in which the gravitational constant is, by definition, constant, and a standard against which the variation of other quantities is judged. Here we can take G, h and c to be constants, and, if we find a variability in the value of the dimensionless number $(G/hc)^{\frac{1}{2}}m$, we can attribute this to a variation in particle masses. Instead of $G \propto \phi^{-1}$ and m constant, we should be able to describe the same theory by G = constant and $m \propto \phi^{-\frac{1}{2}}$.

Dicke (1962a) has shown how this can be achieved by means of a conformal transformation of the metric, by which we mean a new choice of metric $\overline{g}_{\mu\nu}$, related to the original metric, $g_{\mu\nu}$, in which $G \propto \phi^{-1}$, by $\overline{g}_{\mu\nu} = \phi g_{\mu\nu}$. This new metric satisfies the Einstein equations with both the energy momentum tensor of matter and the ϕ-field as the source of curvature, and yields (11.2.1) as the field equation for ϕ. It differs from general relativity in that free particles do not follow geodesics of the metric which solves the field equations. This follows since the equation of motion contains a term involving the variation of the particle mass.

The conformal transformation is permitted, according to Dicke, since there is a certain amount of freedom in the choice of metric for a given space-time. In particular, it is possible to choose a unit of length which varies arbitrarily from point to point, and gives rise to an overall indeterminacy of the scale of the metric. In this way it seems that the theory is superior to general relativity, which appears to require that the same choice of units be made at different points, yet provides no way of comparing the choice of units. For example, if the choice of mass unit is the electron mass, it is claimed that in general relativity this is assumed to be constant from point to point; yet precisely because this is the fundamental unit, there is no way of confirming this constancy.

As Dicke points out, and as we have already indicated, this argument is oversimplified, since we do not measure masses but ratios of masses. In general relativity it is assumed that these ratios remain the same from point to point, since special relativity is to be valid in any freely falling frame, and this can be checked by experiment. In particular the mass ratio $(G/hc)^{\frac{1}{2}}m$ is assumed to be constant and this is open to experimental verification. On the other hand it is possible to use an arbitrary mass unit in general relativity, as in any other theory, provided the correct values of the dimensionless numbers are maintained by compensating changes in the units of length, say. Of course, in

186

this way one gets a different representation of the same theory, usually a much more complicated one. But in this respect there is no difference of principle between the Brans-Dicke theory and general relativity. Nevertheless, they *are* different theories: in order to accommodate variable particle masses in general relativity, one would have to have a variable gravitational constant as well (or the variation of some equivalent quantity); in Brans-Dicke one can, and does, have variable masses alone.

In fact, neither theory is truely invariant under changes of scale; in neither is the conformal factor a matter of convention in the sense that conformally related metrics are indistinguishable and equivalent. Both theories satisfy the strong equivalence principle and this determines, as we have seen, the space-time metric measured by clocks and rods, or, equivalently, the trajectories of free-fall, including the conformal scale in any prescribed system of units. They differ in that Brans-Dicke theory does not satisfy the super-strong equivalence principle, since gravitational effects in a freely falling frame are influenced by a long range interaction with distant matter, whereas general relativity satisfies this version of the principle. Thus, the Brans-Dicke theory differs in postulating a variable strength for the coupling of gravity to matter and in that alone. It should be noted that what Dicke refers to as the strong principle of equivalence, we have called the super-strong form (see Dicke, 1957a,b; 1962b).

How then does a variable coupling to gravity relate to the expression of Machian principles? According to Dicke, through the imposition of Machian boundary conditions on the ϕ-field. Thus, the solution of the wave equation for ϕ is given in terms of a sum over sources propagated over the curved space-time by means of a Green function, or propagator, $G(x,x')$, satisfying

$$\Box\, G(x,x') \;=\; (-g)^{-\frac{1}{2}}\, \delta^4(x,x')$$

where $g = \det g_{\mu\nu}$, and $\delta^4(x,x')$ is the Dirac δ-function in space-time. This yields

$$\phi(x) \;=\; \frac{8\pi}{c^4}\,(3+2\omega)^{-1}\int G(x,x')\,T(x')\,(-g(x'))^{\frac{1}{2}}\,d^4x'$$

if a boundary condition on the solution of (11.2.1) is imposed restricting ϕ to vanish in the absence of sources. Without this condition, empty space solutions ($T_{\mu\nu} = 0$) of the Brans-Dicke equations would be possible, just as is the case for the Einstein equations. One might note, however, that for a universe containing 'matter' in the form of pure electromagnetic fields only, we have $T = 0$, and hence $\phi = 0$ if we impose the Machian condition. Such solutions are therefore impossible. (For a criticism from the point of view of Mach's principle see also: Toton, 1968; 1970).

The theory says nothing about the imposition of suitable boundary conditions on the metric fields to render the behaviour of solutions Machian in character. Despite the implicit hopes of the authors of the theory, this cannot be unnecessary, since just the same problems must arise here as in conventional general relativity with regard to such matters as rotating universes and asymptotically flat solutions. Of course, the theory may nevertheless be correct and the existence of the long range scalar interaction verified by experiment. To date, observations (see, for instance, Norvedt, 1977; Partridge, 1977; Reinhardt and Eichendort, 1977) appear to rule out the Brans-Dicke theory for small values of the parameter ω, although in the last analysis ω can always be increased to give results that agree with those of general relativity to within experimental errors. Direct observation of gravitational radiation, which should be possible in the next decade or so, should be able to establish definitively the existence or non-existence of the scalar component. The implication from observations of orbital changes in the binary pulsar is that general relativity is correct and there is no ϕ-field (Taylor et al., 1979).

§11.3 *The Hoyle-Narlikar Theory*

An alternative approach to the construction of inertial mass through the interaction with the matter in the Universe has been taken by Hoyle and Narlikar (1974; see also: Hoyle and Narlikar 1964, 1965, 1966). The essential ingredient of this approach is the elimination of variables describing the fields in favour of an action-at-a-distance formulation. At its simplest level the idea can be illustrated in terms of Newtonian gravity. Here one can write the force on a body of mass m, due to bodies of mass m_i at distance r_i, directly as

$$\mathit{F} = Gm \sum \frac{m_i}{r_i{}^3} \mathit{r_i} \quad .$$

Therefore, one can construct a potential energy and hence a Hamiltonian or Lagrangian formulation of the dynamics of particles directly in terms of variables relating to the particles. The potential of the gravitational field ϕ, and the Poisson equation that it satisfies, need never be mentioned. Hence the problem of suitable boundary conditions never arises. If one could achieve a similar elemination of the gravitational potentials in a relativistic theory, that is, for example, of the metric coefficients in general relativity, the possibility of arbitrary boundary conditions would not arise, and the dynamics of a body would be manifestly controlled by its interaction with other bodies. In this way Mach's Principle would be incorporated into the theory. Of course, in a Machian theory the proposed relation would have to hold in any reference frame, unlike the Newtonian inverse square law. To achieve this, Hoyle and Narlikar attempt to construct a theory of gravity in which the basic variables are the particle masses obtained by direct interaction with other masses.

The construction of relativistic direct action-at-a-distance theories meets with certain difficulties, since one clearly cannot derive the theory from potential energies depending on the absolutely simultaneous separation of particles. For Maxwellian electrodynamics an appropriate Lagrangian was first given by Tetrode (1922), and the direct particle picture was developed by Wheeler and Feynman (1945, 1949) in their 'absorber theory'. The basic idea here is that the electromagnetic field can be eliminated if for every emission of radiation there is also an absorption somewhere in the Universe. Incorporated into a cosmological model this clearly leads to a selection rule for acceptable cosmologies.

There are several versions of the Hoyle-Narlikar theory; we follow the discussion in Hoyle and Narlikar (1974) with minor changes in notation. A mass function at point x, $m^A(x)$, due to particle A is defined in terms of a Green function $G(x,A)$ which propagates the influence of particle A to the point x. The Green function is chosen to satisfy a wave equation of the form

$$\Box\, G + \frac{1}{6}\, RG \;=\; (-g)^{-\frac{1}{2}}\, \delta$$

where R is the Ricci scalar curvature of the space-time to be constructed from the theory. This form is chosen since it is conformally invariant, a notion we shall discuss in a moment. The mass function is then defined by

$$m^A(x) \;=\; + \lambda^2 \int G(x,A)\, da \quad , \qquad (11.3.1)$$

where $\lambda > 0$ is a coupling constant, and the integration is over the world-line of particle A.

A conformal rescaling of a metric $\underset{\sim}{g}_{\mu\nu}(x)$ to another metric $g_{\mu\nu}(x)$ is a relation of the form $\underset{\sim}{g}_{\mu\nu} = \psi g_{\mu\nu}$, where ψ is a non-vanishing but otherwise arbitrary function of position. An equation is conformally invariant if it maintains the same form, after a rescaling of the dependent variables if necessary, when the metric undergoes a conformal rescaling. For example, the defining equation for the Hoyle-Narlikar Green function is conformally invariant, since the simultaneous replacements $g \rightarrow \tilde{g} = \psi g$, $G \rightarrow \tilde{G} = \psi G$ produces an equation of the same form in terms of the new variables. Hoyle and Narlikar argue for the conformal invariance of their theory on the basis that the other known equations describing zero rest mass fields, namely Maxwell's electrodynamic equations and the Pauli neutrino equation, are conformally invariant, and it is therefore reasonable to attempt to derive a gravitational theory with this invariance. In addition, conformal invariance is supposed to be related to the ability to choose a scale of length arbitrarily, and this is a desirable feature of any theory. We shall return to this second point.

From the mass function, Hoyle and Narlikar are able to construct a variational principle of a particularly simple form

$$\delta \left(- \frac{\lambda^2}{2} \sum_A \sum_B \int \int G(A,B)da db \right) = 0 \quad .$$

This yields the equations of motion of a particle, and can be thought of as giving also the field equations determining the metric out of which the Green function is constructed. Of course, it is not essential in a direct interaction picture to return to the field variables, but one can always do so if desired in order to compare the theory with field theories. The comparison yields the interesting result that for a particular choice of conformal factor in the metric, one such that particle masses are constant, a choice which can always be made, the corresponding field equations for the metric are precisely those of general relativity.

How then does the theory differ from general relativity? If, as Hoyle argues, the choice of conformal frame is arbitrary as far as physics is concerned, we are free to choose it such that the equations are those of general relativity, and hence the solutions are the same. However, the important difference is that only results that have a conformally invariant significance will have any physical status.

For example, the Robertson-Walker metrics are conformally flat (Infeld and Schild, 1945). This is easy to demonstrate for the $k = 0$ model, since we can write

$$ds^2 = a^2(\eta)(-d\eta^2 + dx^2) \quad ,$$

where $\eta = \int dt/R(t)$ and $a(\eta) \equiv R(t(\eta))$. It follows that the sin-

gularity at $t = 0$ arises from a singularity in the conformal
factor $a(\eta)$ at $\eta = 0$.   If the conformal factor is non-physical,
since it can be chosen arbitrarily, so must the singularity be
non-physical. According to the Hoyle-Narlikar theory there is
another half to the Universe before the big-bang (Hoyle, 1975).
        This argument would appear to be incorrect. Just as in
the Brans-Dicke theory, there is a metric which is physically
preferred, since in the freely falling frames of that metric the
laws of physics take on the standard special relativistic form.
Indeed, in the Hoyle-Narlikar theory, even the super-strong principle of equivalence is deemed to be satisfied, since the dimensionless 'constants' of Nature are taken to be genuinely constant;
in the Hoyle-Narlikar theory a variable mass entails also a
variable gravitational 'constant'. Now the theory may certainly
be represented in any conformal frame since, as discussed in
Section 11.2, general relativity may be represented in any conformal frame by an appropriate choice of units. To pass from one
conformal frame to another one redefines the fundamental unit
which, following Hoyle, can be chosen to be the unit of mass.
This redefinition is arbitrary *except* that the unit may not be
chosen to be zero. But it transpires that this is precisely the
choice required at $t = 0$ in order to pass through the singularity.
Thus the singularity has not been abolished but merely transferred
to the choice of unit. Even in general relativity one may add a
previous history to the Universe before the big-bang. The problem
is that no one history is to be preferred over any other, since
the singularity prevents the transmission of information. The
Hoyle-Narlikar theory gives no reason to prefer a conformally related extension.
        To gain further insight, one can contrast the conformal
invariance claimed for the Hoyle-Narlikar theory with the properties of a genuinely conformally invariant theory. For example if
the World were made of gravity and light there would indeed be no
way of defining the conformal factor in the metric by observation.
For the free-falls of photons do not depend on that factor, and
give rise not to an affine connection but to a whole class of conformally related connections. Field equations for such a World
would be constructed so as not to depend on which of the class
were chosen. Of course, there would be no massive particles to
provide a constant unit of mass. One may compare the situation
with that of the passage from Newtonian theory to relativity.
Either theory can be written in arbitrary coordinates, although
Newtonian theory looks simpler if one adapts the coordinates to
the constant absolute time sections. The greater invariance of
relativity arises from the non-existence of absolute time, not

191

from the use of arbitrary coordinates. To pass to a conformally invariant theory it appears that one must abolish the constructs which appear to define for us the privileged conformal frame; that is, it appears that one has to do without massive particles (at a fundamental level, at least), and one has to find truely conformally invariant field equations which do not change in form under conformal transformations.

Another claim made for the Hoyle-Narlikar theory is that it predicts the attractive nature of gravity. This claim would appear to be quite unfounded. For it depends on taking $\lambda^2 > 0$ in (11.3.1); since, however, λ itself never appears unsquared, there is no reason why λ^2 should not be negative. Indeed (11.3.1) is appropriate if all particles in the Universe have the same inertial mass at any point. In reality, one must replace λ^2 by λ_i^2 appropriate to the type of particle under discussion (electron, proton, planet, etc.). Thus λ^2 is not really a coupling constant at all, but a conventional gravitational mass in heavy disguise.

There is one further criticism that may be raised against the Hoyle-Narlikar theory. It has always been a source of difficulty for direct interaction theories that they cannot accommodate the needs of quantum mechanics. For, while the real photons of classical theory may be eliminated, the virtual photons of quantum field theory cannot be expressed in particle terms. The virtual photons appear to arise as fluctuations of the quantum vacuum and ascribe physical reality to the electromagnetic field. The direct particle picture on the other hand requires that the fields should be mere mathematical fictions. Now Hoyle and Narlikar claim to reproduce quantum electrodynamics on the basis of the direct interaction picture, hence to solve this problem. How does this claim stand up? It is difficult to be dogmatic here, but the crucial step in the argument is the division of the vector potential of the electromagnetic field into positive and negative frequency parts. This splitting cannot be carried out in terms of the matter variables. Consequently the electrodynamic quantum vacuum cannot be defined in terms of the matter, and at this point the reality of the field expresses itself (see also Deser and Pirani, 1965).

It may be that a direct interaction picture of the Universe is possible, and if so it may be that Mach's Principle can be incorporated in such a scheme. It appears to us that this objective has not been achieved.

Chapter 12:

Integral Formulation of General Relativity

A theory expressed in terms of a determinate system of linear differential equations and boundary conditions may be stated equivalently in terms of a superposition of the contributions from the sources of the field as an integral representation of the field. Instead of writing a differential operator acting on the field on the left of the equation, one writes the inverse of the operator - a Green 'function' - acting on the sources on the right, together with a contribution from the boundary values of the field. If the sources and the boundary values of the field are known, this provides a solution of the problem. If general relativity were of this form, Mach's Principle could easily be expressed directly by writing the equations for the field as an integral over sources with appropriate boundary conditions explicitly included.

Unfortunately, the Einstein equations are neither linear nor determinate. The lack of linearity means that we certainly cannot express the metric potentials as a linear superposition of known contributions. Remarkably, it turns out that we can implement an idea of Hoyle and Narlikar, and express the metric as a linear superposition of unknown contributions in this sense: that the effect of a distant piece of matter is to be propagated through that curved space-time appropriate to the sum of the contributions of all the matter. Clearly this gives a representation of the metric in terms of a propagator, or Green function, which itself involves the metric. We do not arrive at a solution of Einstein's equations, but an alternative form of the equations. The advantage of this form is that it explicitly contains the boundary values of the metric; in principle one could write into the formula explicitly Machian boundary conditions. This would be the case if the equations were determinate. One would arrive at a definitive expression of Mach's Principle by adopting this appropriate form for the field equations.

However, we have seen that the Einstein equations are indeterminate, involving for their solution the imposition of coordinate conditions and the consequent solution of constraints. The appropriate Mach condition should then be simply that the free wave modes of the field, those not tied to matter, are initially unexcited. It has so far proved impossible to incorporate this directly into the formalism. Instead, an indirect method has been developed whereby a given solution can be tested to see if it satisfies certain criteria, which, it is to be hoped, form an equivalent expression of this Mach condition. Solutions satisfying this requirement are called *Machian*, and Mach's Principle is thereby implemented as a selection rule in general relativity.

§12.1 *A vector theory of inertia*

The full theory has its origins in a model for inertia due to Sciama (1953). The purpose of this model is to show how inertial induction can come about if one has acceleration dependent forces, and how such a result imposes conditions on the material content of the Universe. This is achieved without the technical complications of general relativity in order to isolate the ideas, and to see how they might be implemented in the full theory.

Suppose a force accelerates us relative to the stars. From our point of view the stars are accelerated relative to us. In order to balance the imposed force we say that an inertial force must appear in our frame. According to Mach's Principle this must be due to some acceleration dependent force exerted by the stars. We know of at least one acceleration dependent force, namely the force on a charged body due to electromagnetic radiation from a second, accelerated charge. This is proportional to the product of the charge of the body generating the field with the charge of the body on which it is acting, proportional to the relative acceleration, α, and, to a first approximation, inversely proportional to the spatial separation, r, of the two charges. The direction of the force is parallel to the acceleration. We therefore adopt a model for inertial induction having the same general form, except that charges are replaced by masses. There is a certain degree of justification for this in that the linear approximation to general relativity can be put in the form of Maxwell's equations (Davidson, 1957). The fact that the equations of the theory are formally similar to Maxwell's equations means that this approximate theory will exhibit anti-Machian effects, so the details should not be taken too seriously.

The inertial inductive force on a mass m from each mass M in the Universe is therefore to be written as

$$\frac{Km}{c^2} \frac{M\alpha}{r}$$

where K is a constant and c, the velocity of light, is introduced for later convenience. We shall need to estimate the value of K. This is given by the fact that we know that in Newtonian theory the inertial force is just $m\alpha$. Hence

$$K\Sigma M/r \;=\; c^2 \;. \tag{12.1.1}$$

The sum is given approximately by the ratio of the mass contained in the visible Universe to its radius. We take the Universe to have a density $\sim 10^{-29} g \; cm^{-3}$. This is a lower limit determined by the fact that our failure to detect the local effect of the Sun on a Foucault pendulum means that this effect must be dominated by the contribution to (12.1.1) from the rest of the Universe. It is in fact somewhat higher than the currently popular estimates, but that is of no significance in this crude model. This gives a value of K of approximately $10^{-7} \; cm^3 \; g^{-1} \; sec^{-2}$.

Now the electric force has also a static part proportional to the product of charges and the inverse *square* of the distance. The constant of proportionality is the same except for the removal of factors of c necessary on dimensional grounds. We take a similar static counterpart to the inertial inductive force, and compute the acceleration induced by the Earth because of this force. The acceleration is

$$\frac{KM_\oplus}{R_\oplus^2}$$

where $M_\oplus$ is the mass and $R_\oplus$ the radius of the Earth. Substituting numerical values gives an acceleration of about $10^3 \; cm \; sec^{-2}$, the same for all bodies. This is of the same order as the acceleration due to gravity on the Earth. The static counterpart of the inertial force must therefore be the Newtonian part of the gravitational force, and K, of course, is just the gravitational constant, G,

195

in disguise. Equation (12.1.1) is then just the *sum for inertia* to which we have previously referred (Sections 9.5 and 11.2).

The identity of gravity and inertial forces is already contained in Einstein's equivalence principle. What the model highlights is how these forces must be composed of sums of contributions from the matter in the Universe if Mach's Principle is to be satisfied. The sum of the inertial induction, *1/r*, force derives its main contribution from distant matter, local bodies providing only a negligible addition. For the static $1/r^2$ law the situation is reversed. The contributions from the Universe cancel to zero by symmetry, and the dominant effect comes from local matter, the Earth in the case of terrestrial phenomena, the Sun in the case of planetary motion. In relativity the inertial-gravitational forces are contained in the affine connection or, equivalently, the metric. Consequently, in a Machian theory these must be expressed as sums of contributions from matter in the Universe.

§12.2 *The Integral Form of Einstein's Equations*

In dealing with the full tensor theory the simplest equations we could write would be of the form (Sciama 1964; for review see also: Ellis and Sciama 1972, Sciama 1974):

$$\Box \, \tilde{g}_{\mu\nu} \; = \; -\kappa(-g)^{\frac{1}{2}} \, T_{\mu\nu} \quad ,$$

where $\Box$ is the flat space-time wave operator, $\eta^{\mu\nu} \, \partial^2/\partial x^\mu \partial x^\nu$ in Minkowski coordinates. The conservation laws take the form $\partial/\partial x^\mu \, T^{\mu\nu} = 0$, and for consistency we must satisfy the gauge condition $\partial \tilde{g}^{\mu\nu}/\partial x^\mu = 0$. If $\tilde{g}_{\mu\nu} = (-g)^{\frac{1}{2}} g_{\mu\nu}$ these are just the equations of the linearised theory of gravity, and can be used to provide a first approximation to general relativity for weak gravitational fields. They cannot stand as the final equations of the theory since they do not take into account the contribution of the energy of the gravitational field itself to the metric. We must add to the right hand side of the equations the energy-momentum tensor of the $g_{\mu\nu}$ field, $\underset{1}{\theta}_{\mu\nu}$ say. Then we have

$$\underset{1}{\Box} \, \tilde{g}_{\mu\nu} \; = \; -\kappa(T_{\mu\nu} + \underset{1}{\theta}_{\mu\nu}) \, (-g)^{\frac{1}{2}} \quad .$$

Now the energy of the matter fields in $T_{\mu\nu}$ is *not* conserved, since it may be exchanged with gravitational energy. We have

$$\nabla_{1\mu} T^{\mu\nu} \equiv \frac{\partial}{\partial x^{\mu}} T^{\mu\nu} - \Gamma^{\mu}_{1\mu\lambda} T^{\lambda\nu} - \Gamma^{\nu}_{1\mu\lambda} T^{\mu\lambda} = 0 \; ,$$

where the *covariant derivative*, $\nabla_{1\mu}$, is taken with respect to the metric $g_{1\mu\nu}$. Iteration of this procedure leads to Einstein's equations (Deser 1970).

At each stage of the iteration the equations have the form of a wave operator acting on the unknown field equal to the energy-momentum of the matter and gravitational fields. In order to maintain covariance in the limit, we cannot split off the energy of the gravitational field, but must regard it as part of the left hand side of the equation. This has the unfortunate effect of changing the curved space wave operator, $\Box$, which has an inverse and which would immediately give rise to an integral representation, into a non-linear expression which does not, namely $R_{\mu\nu}(g)$. Nevertheless, one feels it might be possible to carry out some similar procedure which would lead to an integral rather than differential representation of the Einstein equations, since this is clearly possible at each intermediate stage, albeit in a non-covariant way with a coordinate gauge specified by the gauge conditions.

This problem was considered by Al'tschuler (1961) and independently by Lynden-Bell (1967). They showed that the Einstein equations could be expressed in terms of a *linear* operator, depending on the metric of the space-time to be constructed, and acting on the metric of that space-time. This linear operator can be inverted to yield an integral representation of the metric. Unfortunately, the operators obtained by Al'tschuler and by Lynden-Bell were not the same, and the representations are inequivalent in the sense that they involve the splitting between matter and boundary conditions in a different way, and lead to different Machian criteria. However, Gilman (1969; see Sciama et al., 1969) showed that a set of reasonable conditions could be imposed on the representation, which then lead to a unique result. These conditions are (i) that the potential be either the metric itself or a so-called 'density' formed by multiplication by one or more factors of $(-g)^{\frac{1}{2}}$; more complicated possibilities are not to be considered. (ii) the representation is to be *stable* in the sense that the Green function must be appropriate, without modification, to the superposition of perturbations of the matter fields, and yield a representation of the perturbed Einstein equations; this will become clearer below. (iii) the operator is formally

197

self-adjoint (equation 12.2.3);   this gives rise to a well-
defined splitting of the contributions from matter and from the
boundary conditions.

The result is the Sciama-Waylen-Gilman (1969) representation, which we can derive in the following way. We start from
the Einstein equations, considered as equations for $g^{\mu\nu}$ with
contravariant indices. This is important since the procedure
will not work for other choices. The equations are then varied
by substituting $g^{\mu\nu} + \delta g$ for $g^{\mu\nu}$, retaining only terms of first
order in $\delta g^{\mu\nu}$. One obtains

$$\delta R_{\mu\nu} \quad = \quad -\nabla_\lambda \delta\Gamma^\lambda_{\mu\nu} + \nabla_\nu \delta\Gamma^\lambda_{\mu\lambda}$$

where ∇ denotes a covariant derivative,

$$\delta\Gamma^\lambda_{\mu\nu} \quad = \quad \tfrac{1}{2} g^{\lambda\alpha} \{ \nabla_\nu \delta g_{\mu\alpha} + \nabla_\mu \delta g_{\nu\alpha} - \nabla_\alpha \delta g_{\mu\nu} \} \ ,$$

and

$$\delta g_{\alpha\beta} \quad = \quad -g_{\alpha\mu} \, g_{\beta\nu} \, \delta g^{\mu\nu} \ .$$

Writing

$$\delta\psi^\mu \quad = \quad \nabla_\nu (\delta g^{\mu\nu} - \tfrac{1}{2} g_{\alpha\beta} \, \delta g^{\alpha\beta} \, g^{\mu\nu}),$$

we obtain

$$\delta g^{\mu\nu} + 2\nabla^{(\mu} \delta\psi^{\nu)} - 2R^\mu{}_\lambda{}^\nu{}_\rho \, \delta g^{\lambda\rho} \quad = \quad -2\kappa \, \delta K^{\mu\nu} \ ,$$

where

$$K^{\mu\nu} \quad = \quad T^{\mu\nu} - \tfrac{1}{2} g^{\mu\nu} \, T \ .$$

The crucial point now is that, since $\nabla_\rho g_{\mu\nu} = 0$, we can add
twice the Einstein equations to each side to obtain the Sciama-
Waylen-Gilman differential equation:

$$\square\, (g^{\mu\nu} + \delta g^{\mu\nu}) + 2\nabla^{(\mu}(\psi + \delta\psi)^{\nu)} - 2R^\mu{}_\lambda{}^\nu{}_\rho(\delta g^{\lambda\rho} + g^{\lambda\rho}) \ = \ -2\kappa(K^{\mu\nu} + \delta K^{\mu\nu}) \ . \tag{12.2.2}$$

The operator on the left hand side of (12.2.2) is self-adjoint
in the sense that

$$\int uLv \ d^4x \quad = \quad \int vLu \ d^4x \tag{12.2.3}$$

for a self-adjoint operator L. This can be verified by multiplication of (12.2.2) by an arbitrary 2nd rank tensor and integration by parts, neglecting boundary terms. Furthermore, the operator is clearly stable, in the sense of condition (ii) above, since suppressing the variation terms, $\delta g^{\mu\nu}$, does not affect it. This stability property does not follow if we choose variables other than $g^{\mu\nu}$ as basic.

If we let $\delta g^{\mu\nu} \to 0$ we recover Einstein's equations. If, however, we obtain an integral form of the equations, which is possible for a determinate linear system, and then let $\delta g^{\mu\nu} \to 0$, we obtain Einstein's equations in an integral representation. In fact the equations are not determinate. We can perform an infinitesimal coordinate transformation on both sides and the resulting system is unchanged. This can be verified directly by considering the general infinitesimal coordinate transformation, $x^\mu \to x^\mu + \xi^\mu$ for infinitesimal ξ^μ, under which

$$\delta g^{\mu\nu} \to \delta g^{\mu\nu} + 2\nabla^{(\mu}\xi^{\nu)}$$

$$\delta K^{\mu\nu} \to \delta K^{\mu\nu} + LK^{\mu\nu} \quad .$$

Here $LK^{\mu\nu}$ is written for $-\xi^\lambda\nabla_\lambda K^{\mu\nu} + K^{\lambda\nu}\nabla^\mu\xi_\lambda + K^{\mu\lambda}\nabla^\nu\xi_\lambda$, the *Lie derivative* of $K^{\mu\nu}$ along ξ^λ. To invert the system (12.2.2) it is necessary to impose a gauge condition to fix the variation in the coordinates. Note that this does not fix the coordinates themselves, but only how they change in the variation; thus the development is still explicitly covariant.

The obvious choice of gauge condition is

$$\psi^\mu + \delta\psi^\mu = 0 \quad . \tag{12.2.4}$$

This is the analogue of the Hilbert-de Donder condition in the linearised theory, and of the Lorentz gauge condition in electrodynamics. Since we invoke this condition everywhere, and not merely on some initial surface, it is necessary to check that it is compatible with the field equations. There are two ways of doing this. By taking the covariant divergence of the Sciama-Waylen-Gilman equations (12.2.2), using the conservation law for the stress-energy tensor and reordering covariant derivatives, one shows that ψ_μ satisfies a wave equation without sources. Thus once zero ψ_μ will remain zero. Alternatively, and more simply, one shows that from an arbitrary $\overline{g}^{\mu\nu} + \delta\overline{g}^{\mu\nu}$ which does not satisfy $\psi^\mu = 0$, one can always obtain $\psi^\mu = 0$ for a $g^{\mu\nu} + \delta g^{\mu\nu}$ related by a suitably chosen infinitesimal coordinate transformation.

We define now a Green function, $G^{\alpha'\beta'}_{\mu\nu}(x,x')$ which is a second rank tensor at the two points x and x'. The primed labels

199

refer to the point x' and the unprimed to x. In order to accomplish the definition, we need to introduce the parallel propagator A^μ at x to a vector $A^{\alpha'} = \bar{g}^{\alpha'}_{\mu} A^\mu$ at x' by parallel transport. Then $G^{\alpha'\beta'}_{\mu\nu}(x,x')$ is the unique distribution solution of

$$\Box G^{\alpha'\beta'}_{\mu\nu} + 2R^\rho_{\ \mu}{}^\sigma_{\ \nu} G^{\alpha'\beta'}_{\rho\sigma} = \tfrac{1}{2}(-g)^{\frac{1}{2}}(\bar{g}^{\alpha'}_{\mu}\bar{g}^{\beta'}_{\nu} + \bar{g}^{\alpha'}_{\nu}\bar{g}^{\beta'}_{\mu})\delta^{(4)}(x,x')$$

which vanishes outside the past light cone of the point x. by direct computation, or by standard manipulation of Green functions, one can show that the solution to (12.2.2) in the gauge $\psi^\mu = 0$, is

$$g^{\mu\nu} + \delta g^{\mu\nu} = \kappa \int_\Omega G^{\mu\nu}_{\alpha'\beta'}(K^{\alpha'\beta'} + \delta K^{\alpha'\beta'})(-g)^{\frac{1}{2}}d^4x +$$

$$+ \int_{\partial\Omega}\left[\nabla_{\gamma'} G^{\mu\nu}_{\alpha'\beta'}(g^{\alpha'\beta'} + \delta g^{\alpha'\beta'}) - G^{\mu\nu}_{\alpha'\beta'}\nabla_{\gamma'}(g^{\alpha'\beta'} + \delta g^{\alpha'\beta'})\right]$$

$$(-g)^{\frac{1}{2}} dS^{\gamma'} \quad . \tag{12.2.5}$$

The first term is an integration over the volume Ω, the second an integration over the surface bounding this volume, $\partial\Omega$ (Fig. 12.1).

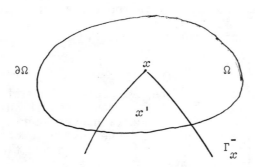

Fig. 12.1. The domain of integration for equations (12.2.5) and (12.2.6).

If we now let $\delta g^{\mu\nu} \to 0$ we obtain a representation of the metric $g^{\mu\nu}(x)$,

$$g^{\mu\nu} = \kappa \int_{\Omega} G^{\mu\nu}_{\alpha'\beta'} K^{\alpha'\beta'} (-g)^{\frac{1}{2}} d^4x + \int_{\partial\Omega} \nabla_{\gamma'} G^{\mu\nu}_{\alpha'\beta'} g^{\alpha'\beta'} (-g)^{\frac{1}{2}} dS^{\gamma'} \quad .$$

$$(12.2.6)$$

Of course, the Green function itself depends on $g^{\mu\nu}$, so this is an equivalent representation of Einstein's equations as integral equations, and not a solution of the equations. The stability property of the Green function can again be seen immediately if we subtract (12.2.4) from (12.2.5), since this then shows the perturbation at x' propagated to x by means of the unvaried Green function $G^{\alpha'\beta'}_{\mu\nu}$. Once the representation has been obtained, we can raise and lower indices in the standard way to give an equivalent integral representation for $g_{\mu\nu}$:

$$g_{\mu\nu} = \kappa \int_{\Omega} G^{\alpha'\beta'}_{\mu\nu} K_{\alpha'\beta'} (-g)^{\frac{1}{2}} d^4x + \int_{\partial\Omega} \nabla_{\gamma'} G^{\alpha'\beta'}_{\mu\nu} g_{\alpha'\beta'} (-g)^{\frac{1}{2}} dS^{\gamma'} \quad .$$

§12.3 *The integral formulation of Mach's Principle*

In fact, at this stage, the indices are required only for technical accuracy and add nothing to the logic of the development, so let us suppress them and write, symbolically

$$g = \kappa \int_{\Omega} GK d\Omega + \int_{\partial\Omega} \nabla G g \, d(\partial\Omega) \quad . \qquad (12.3.1)$$

We may note that (12.3.1) gives a particular meaning to the expression that the metric is *determined by* the matter and initial conditions. For in a non-linear theory the contributions of different bodies do not in general add together, so a specific piece of matter does not determine a specific amount of geometry. Here, however, we see that each element of the stress-energy provides a specific contribution to the metric propagated over the self-consistent space-time geometry.

Since $G = 0$ outside the past light cone of x the integration is in fact over the region of space bounded by this light cone Γ_x^- and $\partial\Omega$. The first term in the representation of g, (12.3.1), the volume integral, gives the contribution to g from material sources inside this space-time volume. The second term, the surface integral, is a solution of the differential equations (12.2.2), (12.2.4), with zero source terms on the right hand side, and represents the contribution to g from the data specified on the intersection of $\partial\Omega$ with Γ_x^-. The data on this surface arises from the contributions of matter at an earlier stage, and from the data for this earlier development. Continuing we see that

201

the surface term represents a contribution from the matter outside Ω together with a contribution from the data at the boundary
of the model universe, which may be at a finite or infinite time
in the past.

In a globally hyperbolic space-time, where the initial
data determines the evolution, we can extend Ω to the whole of
space-time. With this assumption, Mach's Principle may be imposed by requiring the contribution from the boundary of the
space-time to vanish, since the metric is to be determined by
matter. If Ω is taken as the whole of space-time, possibly by
means of some limiting procedure in view of the complications introduced by the existence of singular boundaries, then the surface integral appears to represent only the contribution from the
boundary of space-time. It seems, therefore, that Mach's Principle may be incorporated into general relativity as the requirement that this surface term, taken at the boundary of space-time
must vanish (Gilman, 1970; Lynden-Bell, 1967). On this view,
general relativity is a Machian theory if the field equations are
taken in integral form

$$g = \kappa \int_\Omega G\ K\ d\Omega \ .$$

Gilman (1970) has demonstrated the existence of solutions
to these equations; he shows that they are satisfied for
Robertson-Walker models. On the other hand Minkowski space is
clearly not a solution, since $K = 0$ but $g \neq 0$ in this case. Asymptotically flat solutions are also ruled out. The cosmological
constant is required to be zero, since it could appear in the
theory only as a non-Machian source term. However, we shall find
that this Machian criterion is too strong, since it rules out
solutions in which the metric is, in fact, generated by matter in
a more subtle way.

We have seen that the initial value data for the metric
coefficients are not freely specifiable, but are subject to constraints. In the present case we have imposed the gauge condition
$(\psi^\mu + \delta\psi^\mu) = 0$ on the initial data. In direct analogy with the
electromagnetic case, we can use the gauge conditions to eliminate
second time derivatives from the four $\mu = 0$ equations of the set
(12.2.2) with $\psi^\mu + \delta\psi^\mu = 0$, to regain four constraints on the
initial data. These are just the four $\mu = 0$ equations of the set
(12.2.2) in which gauge conditions have not been imposed. Thus
some of the initial data are obtained by solving the constraints,
and these constrained data may depend on the energy density and

202

flow of matter. Mach's principle does not instruct us to set
these data equal to zero.

   In which cases will the constrained data in fact depend
on the matter distribution? Consider first an asymptotically
flat space generated by a spatially bounded distribution of
matter. As we consider setting the data for the state of this
model universe at successively earlier times, an increasing fraction of the initial universe has a causal influence on a present
observer, since he can see larger domains of initial data come
into view. However, a larger fraction of the visible region is
occupied by empty space and the contribution of the bounded
matter to data increasingly far away becomes relatively unimport-
ant. In the limit as the observer sees an infinite initial space
at t = -∞ the initial data will be essentially independent of the
matter, and will represent purely gravitational energy fed into
the space-time at its 'boundary'. In this case we expect the
Machian theory to impose zero boundary data. Hence asymptotically
flat spaces, if they were to be Machian, would have to satisfy
Gilman's Mach criterion. We see therefore that asymptotically
flat spaces are correctly ruled out by Gilman's condition.

   This can be seen more clearly if we use a technique invented by Penrose (1964, 1968) to represent infinite space-times
by finite pictures. Essentially one applies a conformal rescaling of the space-time which produces infinite compression near
infinity, and one draws a map of the rescaled finite space-time.
This rescaling preserves the distinction between timelike, null
and spacelike surfaces and curves. Applied to Minkowski space-
time, or to asymptotically flat space-time, this procedure yields
the standard Minkowski 'diamond' (Fig. 12.2). One can see that
the metric at P is really determined by data on a null surface,
where the elimination of constraints can be carried out without
difficulty, and the data is essentially freely specifiable.

   Application of a similar rescaling to Robertson-Walker
space-time, with the difference that one infinitely expands the
picture at the singularity at $t = 0$, leads to the 'triangular'
rescaled map of Fig. 12.3. Here one sees that the past light cone
of P intersects the boundary of space-time in a spacelike surface.
Matter outside this surface, which is beyond the horizon of the
observer at P, nevertheless contributes in principle to the data
on the surface. For example, a charged particle at Q makes its
$1/r^2$ electrostatic contribution to the electric field at P even
though no radiation, hence no information, can be transmitted to
P no matter what accelerations are imposed on Q. Ellis and Sciama
(1972) summarise this by saying that one can feel the charge even
if one cannot see it. Clearly one should also be able to feel the

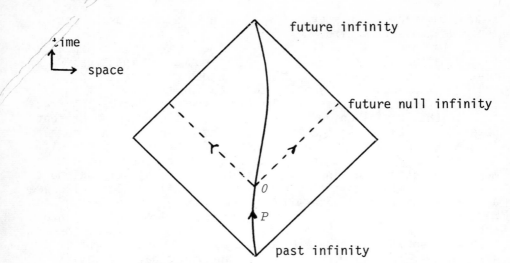

Fig. 12.2. The conformal representation of the (t,r) plane of Minkowski space time, showing the path of a particle P from the infinite past to the infinite future, and the trajectories of light rays emitted at O.

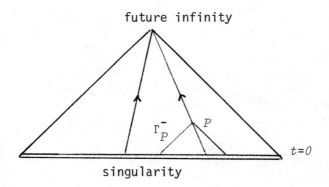

Fig. 12.3. The conformal representation of the (t,r) plane of the $t = 0$ Robertson-Walker model. A charge at Q is outside the past light cone, Γ_P^-, of an observer at P.

contribution of a mass to the inertial-gravitational field. The Gilman condition rules out this possibility and is therefore too restrictive.

Nevertheless the exact Robertson-Walker models satisfy Gilman's condition and are not ruled out! The resolution of this paradox lies in the high symmetry of these models: the particles beyond the horizon contribute to the field, but their contributions sum to zero (Raine, 1971). Suppose, however, one were to introduce an extra neutron into the Universe at Q. This would not affect the propagator, by the stability property, and so would not affect the volume contribution to the metric at P. Its influence must be felt through the surface integral. Gilman's criterion rules out this possibility, and so does not allow the physical fluctuations which are expected to occur.

We are therefore led to attempt to separate the true degrees of freedom in the initial data, and to set the corresponding part of the surface integral equal to zero in order to obtain field equations expressing Mach's principle. So far, attempts to implement this directly have not succeeded and only an indirect approach is available: thus Mach's principle is imposed as a selection rule for solutions satisfying certain conditions which are equivalent, it is hoped, to this direct requirement.

§12.4 *Mach's Principle as a Selection Rule*

The problem is divided into two parts which might be described as kinematical and dynamical (Raine, 1975). In the second (dynamical) part the curvature of space-time, as expressed by the Riemann tensor, is related to the matter currents in the space-time which produce that curvature, by means of a linear superposition of the same general form we have described. Here it is possible to separate the parts of the curvature which may be given as arbitrary initial data from those which must be obtained by the solution of constraint equations, hence determined by matter. This will be dealt with below. In the first (kinematical) part the metric is related to the Riemann tensor by means of a linear superposition.

There is an analogous relation between the vector potential and the electric and magnetic fields in electrodynamics. There one finds that certain parts of the potential are not required to determine the field; for example A^ℓ does not enter the definition of $B = \nabla \wedge A = \nabla \wedge A^t$, and as far as the magnetic field is concerned A^ℓ can be chosen arbitrarily. This is true also of the electric field since only $(A^\ell - \nabla \phi)$ enters E^ℓ and not A^ℓ itself.

If we can isolate parts of the solution of the metric-curvature relation which can be assigned with similar arbitrariness, *over and above the coordinate freedom,* then we should regard such solutions as not giving rise to Machian metrics. For, if the curvature is determined by matter but the metric is not determined by the curvature then the metric is not determined by matter and the solution is not Machian.

The analysis for this first part of the problem proceeds by means of the construction of a stable linear operator which gives the curvature tensor when acting on the metric. This is obtained by variation of the identity

$$g_{\lambda\mu} \, R^{\lambda}_{\nu\rho\sigma} = R_{\mu\nu\rho\sigma}$$

to give

$$\delta g_{\lambda\mu} \, R^{\lambda}_{\nu\rho\sigma} + g_{\lambda\mu} \, \delta R^{\lambda}_{\nu\rho\sigma} = \delta R_{\mu\nu\rho\sigma} \quad .$$

Explicit computation of $\delta R^{\lambda}_{\nu\rho\sigma}$ yields a complicated expression which we can symbolise as

$$L(g + \delta g) = (R + \delta R) \quad . \tag{12.4.1}$$

This yields the Sciama-Waylen-Gilman equation (12.2.2) on taking a trace.

Again the system is indeterminate, as it must be since the coordinates in the varied space can be chosen arbitrarily. If we were simply to impose gauge conditions, we should encounter the same problems as before, since we should have to eliminate constrained variables to set the initial data. However, the system here has one advantage over those we have previously considered, namely that any solution of the source-free equation, by definition, does not contribute to the physical effects of gravity, as measured by geodesic deviation, since it does not contribute to the curvature. Such a solution is therefore either a transformation of gauge or reflects a symmetry of the space-time. Any solution of (12.4.1) which contains a part, γ say, that satisfies the source-free form of equation (12.4.1), (i.e. $L\gamma = 0$), does not give rise to a metric which can be considered as a linear superposition of contributions from curvature as sources, so does not give rise to a Machian space-time.

In order to implement this idea one must provide a distinction between a source-free solution and a solution arising from sources. For otherwise we could simply consider any metric as non-Machian; all we need is to subtract from the supposed

206

Machian metric a solution of the source-free equation, and then add this on again as an explicit anti-Machian component. The unique separation is achieved for a determinate system by means of a Green function, which is an essentially unique inverse of the differential operator. The problem with an indeterminate system is that the operator has no inverse - for otherwise the solution would be determined.

For systems of linear algebraic equations a similar situation may arise. For example, three planes intersect in general in a point, which is the uniquely determined solution of three linear equations. If the planes intersect in a line the system is indeterminate, and any point on the line is a solution. We express this by saying that the system has a *generalised inverse* which provides a set of solutions. For a singular symmetric matrix M, with det $M = 0$, a particular generalised inverse matrix can be thought of as arising by putting M into diagonal form through a suitable change of coordinate axes, and then taking the reciprocals of the non-zero diagonal elements and leaving the zero components as zero. The matrix equation $M\underset{\sim}{x} = \underset{\sim}{f}$ then has a solution $\underset{\sim}{x} = M^I \underset{\sim}{f}$ where M^I denotes the generalised inverse of M. The form of this solution is not unique since $\underset{\sim}{x} = M^I \underset{\sim}{f} + |I - M^I M| \underset{\sim}{y}$ is also a solution for arbitrary $\underset{\sim}{y}$.

We claim that these ideas can be extended to the system of differential equations (12.4.1) to yield a linear representation of $g + \delta g$ in terms of a generalised inverse. If any non-uniqueness of the generalised inverse representation can be absorbed by a different choice of generalised inverse, then we can say that the metric depends linearly on the curvature, and the 'first Mach condition' is satisfied. If an explicit source-free contribution is essential in the representation of g then the first Mach condition is violated.

For example, in Minkowski space-time the metric has the form $\eta_{\mu\nu} = \partial_\mu \xi_\nu + \partial_\nu \xi_\mu$ which, in this case, is just a solution of the source-free equations, since it can be shown that $L\eta = 0$. In Minkowski space-time, of course, the curvature vanishes. Since any generalised inverse of an operator multiplied by zero curvature is zero, the source-free contribution cannot be absorbed by a new choice of the generalised inverse, and Minkowski space-time is not Machian.

Similarly, in asymptotically flat space-times, containing bounded sources, the contribution from the curvature is arbitrarily small at sufficiently large distances from the matter. The approach to a flat metric is provided by a source-free contribution which approaches the Minkowski metric as we go to infinity. These space-times therefore also fail to satisfy the first Mach condition.

207

Another example is provided by plane wave space-times (see Ehlers and Kundt, 1962). The condition $\underset{\xi}{L}R^\lambda{}_{\mu\nu\rho} = 0$ which is satisfied here for ξ^μ not a solution of $\underset{\xi}{L}g = 0$, allows us to construct more than one solution to the system, (12.4.1) differing by a source-free contribution. A plane wave metric is therefore not uniquely determined as a linear superposition of curvature sources, and is therefore non-Machian.

For the dynamical part of the Mach criterion we need a linear relation between the curvature and its sources. For the Ricci part of the curvature obtained by taking a trace, this is provided directly by Einstein's equations. The remaining information in the Riemannian curvature, $R_{\lambda\mu\nu\rho}$, is provided by its trace free part, the so-called Weyl curvature, $C_{\lambda\mu\nu\rho}$. In detail, we have

$$R_{\lambda\mu\nu\rho} = C_{\lambda\mu\nu\rho} + \frac{1}{4}g_{\lceil\mu\mid\rho}R_{\lambda\mid\nu\rceil} + \frac{1}{6}{}_{(}g_{\mid\lambda\lceil\nu}g_{\mu\mid\rho\mid}{}^{)}R . \qquad (12.4.2)$$

The Bianchi identities satisfied by the Riemannian curvature,

$$\nabla_\sigma R_{\lambda\mu\nu\rho} + \nabla_\nu R_{\lambda\mu\nu\sigma} + \nabla_\rho R_{\lambda\mu\sigma\nu} = 0 \qquad (12.4.3)$$

can be considered as differential equations for the Weyl tensor $C_{\lambda\mu\nu\rho}$ by using (12.4.2), and replacing the Ricci tensor terms by the stress-energy tensor using Einstein's equations. Thus we obtain a linear system of equations for the Weyl tensor with derivatives of the stress-energy tensor as sources. In principle this gives rise to a representation of the Weyl curvature, and consequently of the Riemannian curvature, by inversion.

The representation will have two parts, since one is free to add a solution of the equations without sources. If the known curvature of the space-time does not require such an additional term, the space-time will be said to satisfy the second Mach condition, since the curvature is in this case derived entirely from material sources. Otherwise the space-time will be non-Machian, since the curvature contains a source-free contribution.

The mathematical formulation of this condition is unfortunately again beset with certain technical difficulties. In particular the system of equations for the Weyl curvature splits into two types of equations, bearing a strong formal resemblance to Maxwell's equations. On the one hand one has equations involving only derivatives within an initial space-like three-surface in space-time which act as constraints on the initial data. And on the other hand one has evolution equations determining the

208

field off of the initial surface. A method for eliminating the constraints which leads to an explicit formulation of the second Mach condition can be given (Raine, 1975). However, the complexities of the general criterion have so far prevented its explicit use (except in trivial cases), so the details need not concern us here. Applications have been made on the *ad hoc* basis of inspection of the system of equations in particular cases.

We can summarise these results in the following way:

The first Mach condition requires that the metric be determined by the curvature, and hence that local inertial frames are determined by the curvature of space-time.

The second Mach condition requires that curvature be determined by matter, and hence that the deviation of geodesics is determined by the matter content of the Universe.

The sense in which one thing determines another here is specified in the foregoing theory. For both conditions we require the contributions of sources to be propagated through the self-consistent space-time geometry, and to be superposed linearly. Only in this way can one meaningfully describe the contribution of each part of the Universe to the total effect. In a general non-linear theory such a description is not possible. The equations may be soluble for given sources and in this way one may say a solution is determined, but there is usually no way in which the effect of any one part of the source may be described in isolation, since the parts do not add up independently to give the whole. General relativity belongs to a special class of non-linear theories for which a representation as a self-consistent linear superposition is possible.

The origin of self-consistent linear superposition in the Sciama-Waylen-Gilman theory can be clearly seen in an argument due to Lynden-Bell. Let L be the Lagrangian for the Einstein field equations. Then L is a function of the metric $g^{\mu\nu}$ and, suppressing indices which are irrelevant here, we can write the field equations purely symbolically as

$$\frac{\delta L}{\delta g} = T \quad ,$$

since they are obtained by variation of the action with respect to the metric. Here T stands for the energy-momentum tensor. Suppose now that the Lagrangian is homogeneous of degree two in g, an assumption which is satisfied in general relativity provided

209

we take the contravariant metric $g^{\mu\nu}$ as fundamental. Then, by a symbolic form of Euler's theorem,

$$\frac{\delta L}{\delta g} \, g \;=\; 2L \quad .$$

Taking a second variation gives

$$\frac{\delta}{\delta g} \left[\left(\frac{\delta L}{\delta g} \right) g \right] \;\equiv\; \frac{\delta}{\delta g} \left(\frac{\delta L}{\delta g} \right) g + \frac{\delta L}{\delta g} \;=\; 2 \, \frac{\delta L}{\delta g}$$

or

$$\frac{\delta}{\delta g} \left(\frac{\delta L}{\delta g} \right) g \;=\; \frac{\delta L}{\delta g} \;=\; T \; ,$$

which just expresses the homogeneity of degree one of $\frac{\delta L}{\delta g}$. Variation of the field equations leads to equations for the propagation of small disturbances of the form

$$\delta T \;=\; \frac{\delta}{\delta g} \left(\frac{\delta L}{\delta g} \right) \, \delta g \;=\; \frac{\delta}{\delta g} \left(\frac{\delta L}{\delta g} \right) (\delta g + g) - T \quad ,$$

and hence

$$\frac{\delta}{\delta g} \left(\frac{\delta L}{\delta g} \right) (g + \delta g) \;=\; T + \delta T \quad .$$

This is a symbolic 'derivation' of the Sciama-Waylen-Gilman equations which shows that the stable integral representation depends only on the homogeneity of the Lagrangian. It is conceivable that this is an irrelevant coincidence; we have argued that it is this deep structure of general relativity which enables Mach's Principle to be formulated in the theory.

§12.5 *Machian Universes*

In agreement with Gilman's conclusions (Sect. 12.3) and our observations concerning the symmetry of these models, the Robertson-Walker cosmologies turn out to satisfy Mach's Principle. Since there are no symmetries of the Riemann tensor which are not also symmetries of the Robertson-Walker metric, it appears that the first Mach criterion is satisfied - although admittedly no explicit formal proof is known. The second condition is satisfied trivially since the Weyl curvature and the matter currents which give rise to it both vanish.

210

Turning to the simplest generalisation we consider aniso-tropic homogeneous cosmologies of which the Bianchi I models (Sect. 9.5) provide explicitly computable examples. The con-straint equations on the Weyl tensor here reduce to algebraic equations if the initial surface is taken to be one of constant cosmic time, since the curvature of these surfaces is independ-ent of spatial location by the assumption of homogeneity. The Weyl curvature is then conveniently represented by three complex functions, Ψ_j, with $(i = (-1)^{\frac{1}{2}})$

$$\Psi_1 = C_{0101} - \tfrac{1}{2} i(C_{01}{}^{23} - C_{01}{}^{32}) \quad ,$$

and similar expressions for Ψ_2 and Ψ_3. The Bianchi identities may be manipulated to yield three evolution equations for $\underset{\sim}{\Psi} = (\Psi_1, \Psi_2, \Psi_3)$, of the form

$$\frac{\partial}{\partial t} \underset{\sim}{\Psi} + A\underset{\sim}{\Psi} = \underset{\sim}{S} \tag{12.5.1}$$

where A is a matrix involving the metric coefficients, and $\underset{\sim}{S}$ is a vector depending on the energy density of matter. It turns out to be difficult to compute the behaviour of the components of A in terms of cosmic time; but explicit computation is possible if we use a measure of the volume of a given region of the universe as the independent variable. Specifically we use $V = (XYZ)$ where, we recall, the Bianchi I metric is (Section 9.5)

$$ds^2 = - dt^2 + X^2 dx^2 + Y^2 dy^2 + Z^2 dz^2 \quad . \tag{12.5.2}$$

V gives the volume of the space at time t as delineated by partic-ular material particles, clusters of galaxies or atoms, compared with a unit volume at some previous time. Einstein's equations yield $\underset{\sim}{A}$ and $\underset{\sim}{S}$ as a function of V, and V as a function of t.
Suppose the matter content to consist of particles with no proper motions, hence particles which expand exactly, rather than merely on average, along geodesics orthogonal to the homogeneous spatial sections. This is a reasonable approximation to the pres-ent Universe, but not to the earlier stages before galaxies were formed. At earlier times the matter was hot and consequently possessed large random velocities. This case can be dealt with but adds nothing new in principle. In the absence of random motion the mass within a given volume of space expanding with the matter is obviously a constant. This trivial form of the conser-vation law for the energy-momentum content yields the behaviour of $\underset{\sim}{S}$. We find $\underset{\sim}{S} \propto 1/V^2$, and $A \propto 1/V$.

211

Thus (12.5.1) gives

$$\frac{\partial}{\partial V}\, \psi + A_O V^{-1}\, \psi = \rho_O V^{-2} \; .$$

The solution of this equation depending on the source (the particular integral) can be expressed as a series in powers of V starting with V^{-1}. The known Weyl tensor of the space-time computed from the metric (12.5.2) starts with V^{-2}. This is a solution of the homogeneous equation and hence independent of the source ρ. It follows that these models are not Machian.

In principle the procedure for dealing with more complex anisotropic models is the same as in this simple example, although in practice the absence of explicit forms for the matrix A means one must resort to less direct and more technical arguments. The results are however unchanged; all of these models are non-Machian.

The contrast between this result and the Machian character of the Robertson-Walker models is not hard to explain. As we approach the initial singularity in the Robertson-Walker models the dynamical behaviour of the models as specified by the function $R(t)$ approaches that of the Einstein-de-Sitter universe for which $R(t) \propto t^{2/3}$, (with the restriction again that we ignore proper motions). Thus the influence of spatial curvature, specified by the value of k, is irrelevant at early times and the evolution is dominated by a balance between the kinetic and potential energies of matter. In contrast, in the Bianchi, anisotropic models at early times, near the initial singularity, the dynamics is determined primarily by the anisotropy of space and the matter is essentially irrelevant. These models approximate *empty* space solutions initially, and so this failure to satisfy Mach's principle is not surprising.

A similar situation prevails in the case of rotating models. The essence of the proof lies in the fact that rotating, expanding models must also be shearing (Sect. 9.5). The addition of rotation does not affect the approach to empty space at the initial singularity. It follows that the present form of Mach's principle predicts that an exactly spatially homogeneous universe must have zero shear and rotation, and must consequently be exactly Robertson-Walker. Presumably, although it remains to be proved, an approximately spatially homogeneous universe satisfying this form of Mach's principle should have approximately zero shear and rotation. We compare this prediction with observations in the next section. This result is, of course, compatible with Isenberg's discussion (Section 10.3), since 'gravitational energy' does not contribute explicitly as a source in the integral rep-

212

resentation.

Finally, we should note that the above conjecture with regard to approximately spatially homogeneous models is not vacuous: there do exist Machian models which are not homogeneous. An example is the spherically symmetric time-dependent model of Bondi (1947). Here the Universe is represented as an expanding spherically symmetric distribution of particles without random motion, the density varying with radius and time. This is not in fact a satisfactory model of our Universe, since it has a privileged centre, but since, according to the laws of physics, it could have been, it is quite acceptable as an illustrative example. Again we find that in those Bondi models in which the behaviour near the singularity is dominated by matter, the Mach conditions are satisfied, and in the contrary case the second Mach condition is violated.

§12.6 *Comparison with observations*

As a selection rule in general relativity Mach's Principle makes one definite prediction: that a spatially homogeneous Universe should be isotropic. We can conjecture that a small departure from homogeneity will introduce a small anisotropy in a Machian solution. We can therefore ask how well this prediction is confirmed by observations. (For a more detailed treatment see Raine, 1981). The problem has two parts: first, the degree of inhomogeneity of the Universe and second, the degree of anisotropy.

To determine the homogeneity of an instantaneous picture of the Universe at first sight appears to be a difficult undertaking since it seems to require knowledge of what the Universe looks like to different observers at the same time. In fact the inhomogeneity of the matter distribution can be determined by observing the distribution of galaxies over the sky, since a concentration of galaxies in a particular region will appear as a concentration in a certain direction. This has to be distinguished from purely random fluctuations in a system which is homogeneous on average. Thus one counts galaxies as a function of their relative separation and looks to see if galaxies prefer to cluster together on any particular length scale. This would give the degree of 'lumpiness' of the matter distribution. Recent work indicates that there is no preferred scale of clustering and no lumpiness at all on a scale much less than the size of the visible Universe. It appears therefore that the assumption of approximate homogeneity is justified at least for the part of the Universe accessible to observations (see Groth et al., 1977).

In principle it is possible to have a spatially *homogeneous* distribution of *matter* with an *inhomogeneous* spatial *geometry* in such a way that there would be no contradiction with observations at the present time. At later times such a model would depart greatly from the Robertson-Walker cosmology, but there would be no way of deducing this from the present data: one would have to wait and see. It is necessary to assume that our Universe is not of this apparently pathological type, and that the verification of spatial homogeneity of *matter* implies a spatially homogeneous *geometry*.

Direct measurements of the isotropy of the distribution of galaxies is difficult, since we have to measure not only the correlation of galactic positions, but also whether these correlations differ significantly from point to point. There is, fortunately, an indirect method of observation through the microwave background.

This backgroumd, first found by Penzias and Wilson (1965), and Dicke et al. (1965), consists of radiation at centimetric radio wavelengths with a distribution of intensity with respect to wavelengths having the form characteristic of radiation in equilibrium with matter. In the standard big-bang cosmology it is interpreted as the remnant of a hot dense initial phase of the Universe which has cooled as it has expanded. More important for us is that it provides a kinematical inertial frame fixed by the distant matter in the Universe, and can be used to demonstrate the isotropy of the Universe. Observations show that the background radiation is isotropic to a remarkably high precision. This can be expressed by saying that the relative variation in temperature observed over the sky is $\delta T/T \lesssim 0.1\%$. The radiation can appear isotropic only to an observer in one special rest frame, since an observer in relative motion will see a Doppler shift in the frequency of the radiation, which he will measure as a higher temperature in the forward direction and a decrease in temperature in the sky behind him. The observation of only a small variation puts limits on the motion of the Earth relative to the background. Positive detection of an anisotropy has now been reported (Smoot et al., 1977; see also Muller, 1978), and this gives the motion of the Earth as 390 km sec^{-1} in the direction $R.A. \approx 11 \ hrs, \ \delta \approx 6^o$. This is the resultant of a small contribution (~ 30 km sec^{-1}) from the orbital motion of the Earth, a substantial contribution from the motion of the solar system in the Galaxy and an uncertain, but possibly small component due to the motion of the Galaxy in the Local Group. If this last contribution is indeed small there must also be, somewhat surprisingly, either a substantial motion of the Local Group in the Virgo Supercluster, or of the Supercluster as a whole. When this anisotropy is sub-

214

tracted from the data one obtains an upper limit on the temperature variation that could possibly be due to an intrinsic anisotropy in the Universe of $\delta T/T \lesssim 3 \times 10^{-4}$

Note that the ability to measure the motion of the Earth does not contradict special relativity since we are measuring the motion of matter relative to 'matter' (or radiation) in a non-local way, not the motion of matter relative to space. Thus locally one could screen out the radiation, and, in accordance with special relativity, the determination of our velocity could not then be made.

The isotropy of radiation is not yet quite the same as the isotropy of the distribution of clusters of galaxies. However, the radiation we see now has interacted with matter during the earlier hotter phases of the evolution of the Universe, and consequently bears the imprint of the distribution of that hot matter. To a first approximation, we may say that the radiation comes to us from a *surface of last scattering*, which we reach by following rays back into the past until that time (before the formation of galaxies) at which, on average, the radiation last interacted with the free electrons of ionised matter. This surface surrounding us plays essentially the same role as the photosphere of the Sun in providing an approximation to a surface which is to be regarded as the source of radiation. If the matter at this epoch were clumped the last scattering surface would be nearer to us in the clumps than outside them, since the radiation would penetrate the lower density regions with greater ease. Thus the red-shift of the background radiation would very from point to point in the sky and one would observe small scale variations in temperature. This limits the inhomogeneity of the last scattering surface. The details have been investigated by Sachs and Wolfe (1967; see also Rees and Sciama, 1968) who find that the density variations must be less than 10% on the scale of about 10% of the visible Universe.

If the last scattering surface is homogeneous and non-rotating, anisotropy of the geometry will effect the propagation of the radiation, since the Universe is expanding faster in some directions than in others. The characteristic patterns of temperature variation can be computed, and the failure to detect these puts limits on the anisotropy. Stronger limits on anisotropy have recently been obtained by Barrow (1977), in an investigation of the synthesis of other elements from hydrogen in the early Universe. We conclude from this that the visible Universe is spatially homogeneous and isotropic to a surprisingly high precision. On the basis of the Copernican Principle, that our location in the Universe is in no way privileged, one can extend this conclusion

215

beyond our horizons and conclude that the Universe is well represented by a model which departs little from the Robertson-
Walker model. Mach's Principle, as stated here, cannot be used
to explain this completely; but, granted the spatial homogeneity,
our form of Mach's principle predicts isotropy in agreement with
observations.

We can relate the observations of the isotropy of the microwave background more directly to Mach's Principle by considering
rotating spatially homogeneous cosmologies.

If the matter at the surface of last scattering is rotating,
this will add a true Doppler shift to the effects of the dragging
of inertial frames on the propagation of the radiation. The details have been investigated by Collins and Hawking (1973) who
set limits to the possible rotation of the local dynamical inertial frame relative to the local motion of matter. In a homogeneous model one may think of the local motion of matter as providing the kinematical inertial frame. Alternatively, from a
'Newtonian' point of view, one may think of the microwave radiation as propagating freely to us, governed by Maxwell's equations
in a dynamical frame, and carrying information about the rotation
of the distant matter on the surface of last scattering relative
to the local dynamical inertial frame. The results of Collins
and Hawking, which in detail depend on the present density of
matter in the Universe, can be summarised by saying that at no
time in the history of the Universe was the rate of rotation of
the Universe anywhere near as rapid as its rate of expansion. Thus,
at any epoch, the time required for the Universe to rotate once
would be much greater than the time available since the beginning.
Not only, therefore, do the local dynamical and kinematical inertial frames have a relative rotation of less than about 10^{-6}"
per century now, but this small rotation has not been brought
about by the diminution of rotational velocities in an expanding
system, and is not an accident of the time at which we are here
to see the Universe.

To appreciate the precision of this result one can compare
it with the deductions from applications of Newtonian mechanics
(Sciama, 1971). For example, the computation of the dynamics of
the inner solar system is based on a dynamical inertial frame
which rotates more slowly than the outer planets - that is, for
the worst possible case, we take the outer planets to be Mach's
fixed stars. This gives a relative rotation period of less than
about 250 years (the orbital period of Pluto), or about 1^{0} per
year. Continuing the argument, the flattening of the galactic
disc means, according to Newtonian mechanics, that the Galaxy is
rotating with respect to a dynamical inertial frame. Since the

216

Galaxy is in differential rotation with the outer parts moving less rapidly than the inner regions, as the worst possible case, we can take the outer stars to approximate to Mach's fixed stars and to constitute the kinematical inertial frame. Consequently the agreement of the two frames is better than the rotation period of the galaxy or a relative rotation of less than 1" per century. The flattening of the supercluster of galaxies centred on the Virgo cluster, if this were due to rotation, about which there is considerable doubt, could be used similarly to provide a somewhat stronger limit. On purely dimensional grounds, one might expect a rotation rate for the Universe of such an order of magnitude as to give galaxies at the edge of the 'visible' Universe a transverse velocity of the speed of light. Anything larger is ruled out by the absence of transverse Doppler shifts in the spectra of galaxies. This gives a relative rotation of $\sim 1.5 \times 10^{-3}$" per century (Kristian and Sachs, 1966). The limit from the microwave background is considerably more restrictive than any of these limits, and clearly rules out any significant rotation of the Universe.

It is unfortunate that this one firm prediction of the Machian selection rule is a result that is already known, and for which other explanations can be imagined. Thus one might invoke dissipative processes to smooth out initial inhomogeneities or anisotropies. There have been attempts to prove that the work of smoothing has been done by a transfer of energy by photons, neutrinos or shock waves between regions whose density gradients were to be equalised (the so-called mixing processes, see Misner 1968, 1968; Matzner and Misner, 1972). Other suggestions include the quantum mechanical creation of particle-antiparticle pairs in the extremely strong gravitational field near to an anisotropic initial singularity (Zel'dovitch 1972; Zel'dovitch and Novikov, 1975). It appears that this approach could not produce the small departure from isotropy presently observed from arbitrary initial conditions, and, in particular, that exact spatial homogeneity need not lead to exact isotropy.

An alternative approach is a selection rule for model universes based on the requirement that they must admit the evolution of the model builders (Carter 1974; see also Collins and Hawking, 1973). It is, however, difficult to decide the extent to which such a consistency condition requires the Universe to be like the one we observe, and the degree to which such a condition really explains anything. For example, if the luminosity of the Sun were much different from its observed value we should not be here to observe it, but this does not provide a very satisfactory theory of solar structure. The issue devolves round the extent to which

we can talk of other universes in the same way in which we talk
of other suns.

It is possible that much more detailed predictions of the
relation of the departure from homogeneity and isotropy could be
derived on the basis of the Machian selection rule, which could
then form the basis of more rigorous observational tests. One
might note that the Sciama-Waylen-Gilman integral formulation
does not lead to an expression for the *sum for inertia* which could
be used to select particular models, contrary to the hopes of the
original investigations. This arises because the integral formulation is an identity, not an equation. Consider, for simplicity,
the case of solutions, such as Robertson-Walker models, with van-
ishing surface terms. If we take the trace of the representation
(12.3.1) we always have

$$1 \;=\; \kappa \int G^{\alpha'\beta'}_{\mu\nu} \; g^{\mu\nu} \; K_{\alpha'\beta'} d^4x'$$

Thus, for any value of κ, the propagator adjusts itself to yield
the correct sum. We do not therefore select any particular
Robertson-Walker model, and the selection rule admits both spatially closed and spatially open models.

However, one should note that our considerations show that
Mach's principle can have more than a philosophical significance,
and may therefore have a bearing on our understanding of the physical Universe. Certainly, if a rotation of the Universe had
been discovered, Mach's Principle would have been falsified.

Chapter 13: Frontiers of Relativity

> So far, every physical theory of some generality and
> scope ... presupposes for the formulation of its law
> and for its interpretation some space-time geometry,
> and the choice of this geometry predetermines to some
> extent the laws which are supposed to govern the be-
> haviour of matter.   (Ehlers ,1973)

The ancient Greeks searched for a primordial element,
arche, out of which the whole of Nature could be built up. Space-
time may be considered, in a sense, as an arche of contemporary
macroscopic physics. It is not a shapeless *apeiron*, but possesses a skilful internal architecture which has been the subject of
the foregoing chapters. The evolution of physics is, to a large
extent, the evolution of our conception of the structure of space-
time. For the contemporary interpretation this evolution proceeds
towards increasing simplicity and unification; from the distinguished vector fields and points of Aristotelian space-time,
through the separate metric and affine structure of Newtonian
theory with its absolute time; via the unification of geometry
and causality in Minkowski space-time, to the geometrisation of
gravity in the pseudo-Riemannian space-time of Einstein's general
relativity. The guiding principle of this evolution is that the
structure of space-time manifests itself in the behaviour of

219

matter. According to Einstein's interpretation of Mach's doctrine, the space-time structure should be determined entirely by
matter. Yet Einstein's theory fails to achieve this complete
relativity of space-time. What is the response to this failure?
Most physicists would follow Einstein in abandoning the dogma of
an excessive positivism in favour of the theory of relativity as
it stands, since, in the final analysis, a successful physical
theory has no need of a prior philosophy. Fundamentally, evolutionary progress has been, and will continue to be, the fruit of
experiment and observation, not of speculation. But one cannot
interpret an observation *ex nihilo*. It is necessary to know what
the possibilities are. And it is in influencing the development
of these possibilities that the analysis of space-time structure
in general, and of Mach's principle in particular, might continue
to play some role.

§13.1 *Unified Field Theories*

In the context of the physics of the first quarter of the
twentieth century, the successful elimination of the gravitational
force in favour of space-time geometry resulted in a dichotomy in
the description of physics. On one side of the Einstein field
equations stands a purely geometrical quantity, and one the other
the stress-energy density, which must result from what at the time
was the only other known interaction of matter, the electromagnetic force. From a mathematical point of view the theories of
electrodynamics and gravitation have a certain amount in common,
beginning with the inverse square laws of electrostatics and Newtonian gravity. In terms of the overall momentum of the historical development, it was natural for Einstein to endeavour to remove the theoretical dichotomy by including the electromagnetic
force in a geometrical theory of the unified field.
Besides the mathematical representation of the theories,
electrodynamics and gravity have little in common. To develop a
unified geometrical description in the absence of some guiding
physical principle, one must proceed to try all the mathematical
possibilities. Historically this has not proved to be a very
productive way of discovering new theories, but it is not for that
reason necessarily doomed to failure. For example, had general
relativity not been discovered as early and as unexpectedly as it
was, and had quantum theory then dominated physics completely,
the classical theory of general relativity might have been discovered through quantum mechanical arguments by just this method of
exhausting all the possibilities for a self-coupled spin two
theory (Boulware and Deser, 1975). However, in the search for a

unification of gravity and electromagnetism the obvious generalisations of general relativity turned out to be a legion. None
of these were clearly to be rejected, and most of them were hope-
lessly difficult to develop mathematically. For example, one
might consider a 'geometry' based on a quartic invariant interval
$ds^2 = (a_{\mu\nu\rho\sigma}dx^\mu dx^\nu dx^\rho dx^\sigma)^{\frac{1}{2}}$, or on a path-dependent concept of
length, or on a five-dimensional space-time (see Tonnelat, 1955,
1966), and so on. With no clear physical motivation, the enterprise foundered amidst waves of tensor indices.

One possibility that was extensively investigated by Einstein (1955), Schrödinger (1963) and many others, is the introduction of a non-symmetric metric, or a non-symmetric connection.
Thus one has, for example,

$$\Gamma^\lambda_{\mu\nu} = \Gamma^\lambda_{(\mu\nu)} + \Gamma^\lambda_{|\mu\nu|} \quad \text{where} \quad \Gamma^\lambda_{(\mu\nu)} = \Gamma^\lambda_{(\nu\mu)}$$

is symmetric in its indices and $\qquad \Gamma^\lambda_{|\mu\nu|} = -\Gamma^\lambda_{|\nu\mu|}$

is antisymmetric. The paths of neutral test particles are unchanged, since these depend only on the symmetric part of the
connection, but one has clearly introduced an additional antisymmetric tensor field which is not determined by the metric. The
mathematical development of the theory of a non-symmetric connection is due to Cartan (1922, 1923). The antisymmetric part is
referred to as the torsion tensor, and is a standard part of
modern differential geometry. It appears, however, to have no
relation to electrodynamics.

An appropriate physical interpretation of torsion was given
independently by Kibble (1961) and Sciama (1962). They found that
it could represent an intrinsic *spin* of space-time, and could be
generated by the spin-density of matter. To visualise the meaning
of spin at a point consider a spinning body of finite extension,
r, and angular velocity, ω, and imagine the body to be shrunk to
a point, $r \to 0$, in such a way that the angular momentum per unit
mass, $s = \omega r^2$, remains constant (hence $\omega \to \infty$). The spin at the
point is s. It is known that fundamental particles and nuclei
possess spin, and hence give rise to a spin density of matter.
Ordinarily this is zero on average since the spins are oriented
at random. Interest in the torsion theory lapsed when it was
thought that there were no physical circumstances in which the
matter spins could be lined up and sufficiently densely packed to
give a significant contribution to the torsion. In practice it
appeared that torsion could be neglected even if it was felt to
be necessary in principle. The theory has, however, been re-

investigated recently by Trautman and co-workers and may lead to important qualitative effects. For reviews and references see Kuchowicz (1975, 1976) and Hehl (1973, 1974).

An alternative approach to unified field theory was pointed out by Rainich (1924). This consists essentially of simply reading the Einstein equations backwards. From the geometry one reads off the energy-momentum density; from this one extracts the values of the electric and magnetic fields, which automatically satisfy Maxwell's equations. For this reason the theory may be described as *already unified*. This is another aspect of the fact that the Einstein field equations determine the equations of motion of whatever is generating gravitational forces. The result is unique, however, only if electromagnetism alone is considered. Even apart from matter, if there are other forms of radiation present, such as neutrinos, it is now known that the determination of the fields by the geometry is no longer unique.

The theory was discovered independently by Misner and Wheeler (1957), and developed particularly in the direction of a theory of matter as the topology of 'empty' space-time (Misner, 1960). The attempt to build the World out of empty space-time is, of course, the complete antithesis of the Machian view. One can look at it as a 'physical monism' in the Leibnitzian spirit, and as an implementation of Spinoza's doctrine, which also influenced Einstein in the development of general relativity, that the fundamental description of a physical reality should be necessarily closed.

In Wheeler's development of the already unified theory one regards the Einstein-Rosen 'bridge', which joins the two asymptotically flat regions of the Schwarzschild geometry (Sect. 9.2) as a bridge, or *wormhole*, between two distinct parts of one and the same asymptotically flat space-time. Threading the wormhole with electric lines of force causes it to behave, for an outside observer, as if there were charges of opposite polarity located at its ends. Looked at in detail there is no source for the charge, only 'empty' space-time. From the conventional point of view, the space-time is not completely empty since there is a source-free electric field; from the 'already unified' viewpoint this electric field is to be read out of the geometry and not separately introduced.

As models for massive but uncharged particles, Wheeler introduced the idea of geons (Sect. 10.3). These may be regarded as concentrations of electromagnetic or gravitational energy held together by their own gravity. Actually such objects cannot be completely unstable; their lifetime can be long if they are sufficiently large, but they do not appear to correspond to anything at all in Nature.

It is probably not unfair to say that positive results of this theory are hard to find. In this sense one might consider the theory as not 'already unified' but as 'still programmed'. The methodology of the theory has been discussed extensively by Graves (1971).

With the discovery of further forces in Nature, in addition to electromagnetism, the 'strong' or 'nuclear' force which binds the neutron and proton in the atomic nucleus, and the 'weak' force responsible for the beta-decay of the free neutron into a proton, the need to combine only electromagnetism and gravity seemed less urgent. With the development of quantum theory the dichotomy between gravity and electricity was overshadowed by the dichotomy between the non-quantum theory of gravity and the quantum theory of everything else. Einstein hoped that a unified field theory might provide an alternative to quantum physics, the probabilistic nature of which repelled him. Today, progress is being made towards a unified theory, but it is being made in the context of quantum physics (Sect. 13.3).

§13.2 *Machian Theories*

At best, Mach's principle has been expressed in general relativity as a selection rule. To go beyond this and incorporate the priority of matter over space-time directly into physical theory one may have to give up general relativity as the correct or complete description of gravity at the non-quantum level. From this point of view, the weakest part of Einstein's theory is the field equations. Einstein himself made the first unsuccessful attempt to give these an automatically Machian form with the introduction of the cosmological term. Other forms of the field equations have been proposed for various reasons. One can obtain an infinite variety in a Lagrangian approach simply by choosing curvature scalars, other than the Ricci scalar, as a Lagrangian density. The theory obtained must agree with Newtonian gravity in the appropriate limit, and this is a problem in this approach. If one requires formal agreement, this essentially limits the choice uniquely to general relativity. If one requires only agreement in the results obtained for certain situations, in which experimental results and observations are available, it is necessary to calculate each theory in considerable detail. Goldoni (1976) has shown that other methods of generalisation of Newtonian theory are possible. As further support for these investigations, the non-renormalisability of general relativistic quantum field theory, to which we refer below (Sect. 13.3), may indicate that the wrong field equations are being quantised, although it is at least equally likely that the methods of calculation are at fault. Einstein's equations arise very naturally from very little input: essentially the demands of covariance and a 'standard' Hamiltonian formulation. Alternatives are therefore likely to be rather complicated and contrived.

Mach's Principle requires a very deep connection between local physics and the large scale structure of the Universe. Yet in general relativity only at the very last step, the integration of the field equations, do global considerations enter an otherwise purely local theory. It may be that the fundamental assumption which general relativity shares with all our present physical theories, namely that local physics is essentially independent of our global environment, is invalid. One could conceive the possibility that our local physics as explored in a freely falling frame would actually be different if the Universe as a whole were different, or if we inhabited a different type of region of the Universe, if such exists. The introduction of a dependence of inertial mass on the global matter distribution is a development of this kind (Chapter 11). One might consider also theories which make use of preferred vector fields provided by the observed matter distribution such as can be derived from the existence of homogeneous surfaces of constant cosmic time. These might be built into the theory in such a way that the local laws of physics in an inhomogeneous region of space would differ slightly from those we observe, and that the physics in a universe very different from our own would differ completely. One can imagine that it might then be possible to satisfy Mach's principle in all of these possible universes, and that non-Machian universes would be automatically impossible. Care is required to avoid conflict with observations here. We know already from the Hughes-Drever experiment that preferred frame effects linked to the distribution of matter are exceedingly small. Other experiments such as those of Michelson and Morley and of Turner and Hill tell us that preferred velocity effects are likewise small. Of course, one always has complete freedom to introduce into a successful theory parameters which take the value unity in our Universe but are nevertheless determined by the matter distribution and would take different values in other worlds. This unrestrained flexibility in the development of complexity has never proved of much value in the history of science.

An alternative approach to the elimination of space-time as an independent element of the theory, and hence to the implementation of Machian ideas, has been proposed by Barbour (1974). If one uses the coordinates of bodies as the basic dynamical variables, then one can express the motion of a single body in an otherwise empty universe. According to Barbour this is the fundamental reason for the appearance of inertia relative to space in dynamical theories. Barbour therefore uses the relative distances of material particles, instead of coordinates, in the dynamical equations of motion. In terms of relative configuration, the rigid rotation of a system of particles relative to space cannot even be defined; only the motion of matter relative to matter can enter the dynamics from the beginning. The history of the World in this picture is a path in a relative configuration space, and the dynamical variables refer only to that path. As presently construed the theory involves an absolute time, since the Lagrangian is constructed from the simultaneous relative positions of particles. It is not clear that this in itself involves insuperable difficulties. On the other hand, it does allow the construction of what is effectively an action-at-a-distance interaction, and this can mask the necessity for field equations, which may introduce a non-Machian element in any relativistic development of the theory (Barbour 1974; Barbour and Berlotti 1977). It is interesting that the idea may be capable of extension to quantum mechanics (Jones 1974, unpublished): one would introduce, from the beginning, a many-particle action in a summation-over-paths approach, but one would sum only over relative configurations.

§13.3 *Quantum Gravity*

Einstein's theory of gravity predicts the existence of gravitational radiation. In common with all other forms of energy it appears that such radiation must be subject to the laws of quantum physics, since it must interact with quantised matter. It is natural therefore to attempt to apply the procedures of quantum field theory to Einstein's field equations, treating the metric, or some equivalent field, as the basic object to be quantised. This might be labelled the conventional approach to the quantisation of gravity, and one might call the result a general relativistic quantum field theory. It has been the subject of periodic waves of activity over the last fifty years or so.
    Precisely because of the problem of identifying freely specifiable parts of the metric, which we discussed in Sect. 10.2, hence of identifying the variables to quantise, the construction of a theory by this approach met with extreme difficulties. A possible solution to the problem was first given by DeWitt (1964, 1967), who showed how to compute the quantum probabilities of processes involving gravitons in a covariant theory. The computa-

tional rules are derived through a Feynman *sum over histories* (see Feynman and Hibbs, 1965). In this method one obtains the probability for the system to go from one state to another by summing amplitudes for the system to take possible paths between the two given states in the space of states of the system. In the classical (i.e. non-quantum) theory only the one path permitted by the dynamical laws is relevant, since the system will certainly take this path. In quantum theory all paths make some contribution as a reflection of quantum uncertainty. The problem of carrying out this summation in general relativity arises from the restriction that one must add only physically distinct paths, not different coordinate representations of the same path. An elegant solution was proposed by Fadeev and Popov (1967), but is now known to involve a false assumption.

The resulting theory appears to suffer from the appearance of infinite quantities which cannot be removed. Technically, the theory is apparently non-renormalisable, at least if one considers non-empty space-times and in the absence of seemingly unlikely accidental cancellations with contributions which have not yet been calculated. In other quantum field theories, such as quantum electrodynamics, infinites appear which can be removed by the process of renormalisation, or redefinition, of certain constants in the theory. The rationale of this procedure is that the calculations are made on the basis of a theory containing constants which are defined in the absence of interactions, whereas the values are measured in the presence of interactions. The process of renormalisation takes one from the hypothetical values to the measured ones. The theory is renormalisable if the removal of infinities can be accomplished without the introduction of an infinite number of constants. Otherwise the theory is non-renormalisable and is powerless to make predictions; for an infinite number of observations are required simply to fix the values of the parameters of the theory.

The non-renormalisability of general relativistic quantum field theory, if substantiated, is an ironic result in view of the widely held hope that the inclusion of gravity in field theory would remove infinities all together. This would certainly be highly desirable since we have glossed over the inelegant feature of even renormalisable theories that finite values for measured parameters are possible only if the (unmeasurable) unrenormalised values are infinite. The reason for invoking gravity in this context is the belief that a quantised metric, and hence an uncertainty principle for the metric and the extrinsic curvature, will lead to a smearing out of light cones, the positions of which would not be defined with certainty. For zero mass particles at least, it

226

is just the propagation on a fixed light cone that leads to infinities. Consequently the smearing of the light cones might lead to the elimination of these infinities.

We now see that gravity appears to worsen the problems. For the obvious response to non-renormalisability is the abandonment of a useless theory. Thus it might be that Einstein's equations are not the appropriate starting point for quantisation. This would be the case if they turned out to be purely phenomenological equations which provided a model for the large scale structure of space-time where they have been subjected to empirical test, but did not correctly describe the microscopic nature of space-time. Analogously, one does not discover the quantum nature of matter by quantising the equations of the theory of elasticity of solid bodies. Such arguments lead to the more radical approaches to quantum gravity of section 13.4.

An alternative response to non-renormalisability is to attribute it to the feature of the method of calculation which consists of a perturbation expansion approach. Indications of this have been obtained by Hawking (1979). Hawking argues that included in the summation over histories should be contributions from space-times with different topologies, in particular, having different numbers of asymptotically flat regions. One expects quantum gravitational effects to occur on a length scale of the Planck length, $(Gh/c^3)^{\frac{1}{2}} \sim 10^{-33}$ cm, since this is the only length that can be constructed out of the constants of gravity, special relativity and quantum theory. Hawking's results produce fluctuations in the topology of space-time on just this scale. This possibility has been extensively promoted by Wheeler (1964, 1968, 1973). It would lead to a picture of a smooth classical space-time with a detailed quantum microstructure within the framework of a manifold model.

§13.4 *Beyond the Space-Time Manifold*

The theories we have discussed from Aristotle to Einstein and to the development of quantum gravity have one element at least in common. They are theories of the structure to be found or imposed on the space-time manifold. Their common ancestry is the resolution of Zeno's paradoxes through the assumption of a differentiable manifold model for space-time.

It can be argued that the manifold model possesses intrinsically anti-Machian aspects (Heller, 1970). For one may always induce a metric structure in the local neighbourhood of a point from a Euclidean or Lorentz (pseudo-Euclidean) structure of the tangent space at the point. This is achieved by mapping the straight lines in the tangent space into the manifold and defining the geometry to be that for which the mapped curves are geodesics. There is a complete arbitrariness in this mapping, so no unique metric structure is obtained. However, the possibility of this metric structure is inherent in the manifold concept, and is quite independent of whether there is any matter in the space-time. In this sense one can think of it as an absolute element. The role of Mach's Principle, then, can be only to *select* from amongst all possible metrics certain allowed ones determined in some way by the matter content of the particular space-time. From this point of view, once the manifold model is accepted, Mach's Principle can be implemented only as a selection rule.

Another aspect of the inherently anti-Machian nature of the differentiable manifold is the local topological and differential structure which is imposed by the very definition of the manifold without reference to the matter content. We have suggested that this structure is the natural one for the description of the motion of material bodies, but the objection to this gains a certain weight if one takes into account the fundamental quantum nature of matter. The objective reality of a well-defined differentiable trajectory then loses its significance.

On this view the incorporation of quantum theory into space-time geometry would take place at a deeper level than the quantisation of the metric. Wheeler (1964) has argued for a quantisation of topology. Penrose (1968, 1975; Penrose and Mac-Callum, 1973) has advocated that the space-time itself should be linked with the quantisation: in his development of twistor theory the idea is that the points of space-time should be smeared out while the light-cone structure remains well-defined.

Underlying these ideas is the thought that the space-time we employ at a classical level should itself be the result of some smooth averaging procedure, rather than the fixed background on which quantum effects are to be displayed. In his spin-network theory Penrose (1975) constructs a (non-relativistic) Euclidean space from the lattice of relations provided by the relative spins of objects, which may be represented as point particles for the purpose of visualisation. Certain purely combinatorial rules are postulated to enable the probability for various spin values to be computed when an additional particle is added to the network. From this one builds up physical geometry by associating directions with the spin axes of systems with large spin values. The attempt to incorporate time into the geometry, and to define spatial displacement as well as directions, led to the development of twistor theory. The hope is that by concentrating on space-time structure the role of gravity will emerge in some natural way.

228

An alternative approach to some form of lattice geometry might be based on the transformation between discrete and continuous systems which can be accomplished by the so-called *renormalisation group*. The standard example is a system of interacting spinning particles in a solid, which might be postulated to serve as a model for ferromagnetism. By looking at these spins on a large scale, one can attempt to isolate certain characteristic universal features independent of the precise details of the interactions. This model can be transformed in this way into a field theory model in which the general properties are derived not from a system of discrete particles but from a continuous field theory. One might hope to extend this in some way to the structure of space-time itself.

*

For many years after its inception the general theory of relativity could be regarded as standing apart from the rest of physics. Except for some very detailed observations of the planet Mercury and the bending of starlight by the Sun, which had been fully worked out anyway, and some mere speculations on the nature of the Universe as a mathematical whole, general relativity made no contact with either observation or theory. This is now no longer the case. With the rapid advance of technology, and the accumulation of knowledge of the astronomical Universe that this has brought, general relativity is becoming an active tool in the service of our understanding of the physical World. One might think then that quantum gravity, concentrating, as it does, on the structure of space on a scale of 10^{-33} cm, and of time in a moment of 10^{-43} seconds duration, is a pure theoretical luxury unrelated to the development of the rest of physics. This too is probably false: the limits of the classical theory are reached in our present knowledge of the World, and the limits of quantum physics without gravity may already have been transcended.

Already we know that the introduction of quantum mechanics into the space-time of a black-hole leads to behaviour qualitatively different from the classical theory. The mass-energy of the black hole is radiated away in the form of particle-antiparticle pairs and as radiation corresponding to a system with a temperature inversely proportional to the mass of the hole. Not only does a black hole bathed in radiation satisfy the laws of equilibrium thermodynamics, but allowed to radiate freely into space it obeys the fluctuation-dissipation theorem of non-equilibrium thermodynamics. For the macroscopic black holes we believe to exist, the process of evaporation is too slow to be of astrophysical relevance.

But the results provide an insight into the depths of unification still to be achieved (see Hawking, 1979).

In the early stages of the Universe we know we need a quantum theory of matter which will apply in a highly curve space-time where geodesics deviate greatly over the size of an elementary particle. No definitive theory is available here, although we may suppose that particle-antiparticle pairs will be created out of the curvature of space-time. At an even earlier stage we need to take account of the quantisation of gravity. According to classical general relativity, a universe having the general properties we observe ours to have must contain a singularity of some sort, beyond which space-time cannot be continued. In the simplest models, the Robertson-Walker cosmologies, this singularity is a region of infinite curvature in the past. We may assume that before the curvature becomes infinite the limits of the classical theory will be reached and a quantum theory will be needed.

Here then, quantum gravity in some form is needed not only to unify our modes of thought, not only to calculate beyond the limits of our present theories, but to satisfy us that we know how to calculate at all. For through the singularity theorems, general relativity predicts its own inadequacy. It is not merely the inadequacy of a macroscopic electrodynamics which proved unable to deal with the structure of the atom, or of a mechanics which could not deal with bodies moving near the speed of light, but an inadequacy which arises from the prediction of the breakdown of the very ideas upon which the theory is based, the breakdown of the manifold of space-time. The history of our understanding of space-time begins with the paradoxes of Zeno. Out of those paradoxes arises the space-time manifold description of physics. We may assume that the evolution of the relativity of space-time does not end with the paradox of the singularities.

Appendix: Mathematical Notes

The purpose of this note is to provide a brief account of the notation we use. Details can be found in the references to Chapter 1.

Before relativity one would always use coordinates with a metrical significance to specify and compute physical situations. Freeing himself from this metrical association was, by Einstein's own admission, one of the hardest steps towards general relativity. For in general relativity the metrical significance of the coordinates can be apparent only after the metric is determined, that is, after the field equations have been solved. One is therefore forced to work with coordinates which are completely general. To extract physical information it may then be necessary to specialise the coordinates. It is therefore necessary to know whether the results depend on the choice of coordinates, in which case they may be of dubious physical significance unless that choice is itself determined physically, or if the results are independent of the coordinates and therefore represent physically meaningful conclusions. For the discussion of general laws one must require that the relevant equations have the same form in all coordinate systems. To satisfy this requirement equations must relate quantities which transform in the same way when the coordinates are changed. It follows that the way that mathematical quantities change on transformation of coordinates becomes a sign of their physical significance and an important means of classification.

Let $(x^i; \; i = 1,2,3)$ label points in a three-dimensional manifold (the extension to n-dimensions is obvious). The expression for the distance between infinitesimally separated points (x^i), $(x^i + dx^i)$ is

$$ds^2 = g_{ij}(x) \; dx^i \; dx^j \; . \tag{A1}$$

In this equation and throughout the book we employ the *summation convention*, that repeated indices are to be summed over. Thus, explicitly (A1) is

$$ds^2 = g_{11}(dx^1)^2 + g_{12}dx^1dx^2 + g_{13}dx^1dx^3 + g_{21}dx^2dx^1 + \ldots\ldots$$

The significance of the positions of the indices will be revealed presently. The line-element (A1), (with specific functions $g_{ij}(x)$) defines the meaning of the coordinates by associating a distance with pairs of labels.

Now change to a new system of labels $y^\ell = y^\ell(x)$, with the inverse relation $x^i = x^i(y)$. Then, using the summation convention

$$dy^\ell = \frac{\partial y^\ell}{\partial x^i} \; dx^i \tag{A2}$$

and

$$ds^2 = g_{ij}(x) \; \frac{\partial x^i}{\partial y^\ell} \; \frac{\partial x^j}{\partial y^\kappa} \; dy^\ell \; dy^\kappa \equiv g_{\ell\kappa}(y) \; dy^\ell dy^\kappa \; . \tag{A3}$$

From (A3) we see

$$g_{\ell\kappa}(y) = g_{ij}(x) \; \frac{\partial x^i}{\partial y^\ell} \; \frac{\partial x^j}{\partial y^\kappa} \; . \tag{A4}$$

Thus (A4) gives us the rule for changing the metric coefficients when we change coordinates. An object which transforms in this way is called (by definition) a second rank covariant tensor (second rank because of the two indices). An object, call it A^ℓ, which transforms like (A2), viz.

$$A^\ell(y) = \frac{\partial y^\ell}{\partial x^i} \; A^i(x) \tag{A5}$$

231

is called a contravariant vector (= first rank contravariant tensor). The significance of the positions of the indices is that they tell us which way up to put the partial derivatives in the transformation law (A4, A5). The difference is labelled by calling the objects covariant or contravariant.

We define $g^{ij}(x)$ as the inverse matrix to $g_{jk}(x)$. Thus

$$g^{ij}(x)\ g_{jk}(x)\ =\ \delta^i_k \tag{A6}$$

where δ^i_k is a second-rank mixed tensor defined by $\delta^i_k = 1$ if $i = k$, $\delta^i_k = 0$ if $i \neq k$. In matrix language, (δ^i_k) is just the identity matrix. One can show that the index positions on $g^{ij}(x)$ correctly specify the transformation law. One can then show that from A^ℓ one can construct $g_{k\ell}A^\ell$ which transforms like a covariant vector; that given a covariant vector B_k, the object $g^{\ell k}B_k$ transforms like a contravariant vector; and that $g^{ik}g_{k\ell}A^\ell = A^i$. Hence we define $A_k = g_{k\ell}A^\ell$, with an obvious extension to objects with more than one index. This is consistent with (A6). The point of this discussion is that A_k and A^ℓ are not essentially different objects, but different representations of the same object. In Euclidean geometry, where g_{ij} and its inverse are themselves the identity matrix there is clearly no numerical difference between A_k and A^ℓ, and the distinction can be dropped.

In particular cases the notion of a tensor and the defining transformation law may be specialised. We have just noted one such example, namely Euclidean geometry. Here we can restrict ourselves to coordinate systems in which ds^2 takes the usual simple form, $dx^i dx_i \equiv (dx^1)^2 + (dx^2)^2 + (dx^3)^2$ and the only transformations allowed are rotations and translation which preserve this form. In that case the corresponding physical objects are called Cartesian tensors. The notions advanced here are completely consistent with the idea of a Cartesian vector as a directed line-segment. In this case we also use the notation $\underline{A}$ for the vector.

A second example is Minkowskian geometry. Here we take a coordinate system $(x^\alpha, \alpha = 0,1,2,3)$ with ds^2 defined by the usual form, $ds^2 = dx_\alpha dx^\alpha \equiv -(dx^0)^2 + (dx^1)^2 + (dx^2)^2 + (dx^3)^2 \equiv \eta_{\alpha\beta}dx^\alpha dx^\beta$. The metric form is preserved by Lorentz transformations, and Lorentzian tensors transform in the appropriate way under this restricted set of transformations. Here there is a difference between covariant and contravariant components: e.g. $(A_\mu) = (-A^0, A_1, A_2, A_3)$. A Lorentzian vector is often called a 'four-vector'.

Note that we adhere to the convention that Latin indices i,j,k ... range over $1,2,3$ and Greek indices, α, β ... μ, ν ... over $0,1,2,3$.

We are now in a position to state the Einstein field equations. The affine connection associated with a metric $g_{\mu\nu}(x)$ is given by

$$\Gamma^\lambda_{\mu\nu}\ =\ \tfrac{1}{2}g^{\lambda\alpha}\left[\frac{\partial g_{\alpha\mu}}{\partial x^\nu} + \frac{\partial g_{\alpha\nu}}{\partial x^\mu} - \frac{\partial g_{\mu\nu}}{\partial x^\alpha}\right]\ .$$

The Riemannian curvature corresponding to this connection is given by

$$R^\lambda{}_{\mu\nu\rho}\ =\ -\frac{\partial\Gamma^\lambda_{\mu\nu}}{\partial x^\rho} + \frac{\partial\Gamma^\lambda_{\mu\rho}}{\partial x^\nu} - \Gamma^\lambda_{\sigma\rho}\Gamma^\sigma_{\mu\nu} + \Gamma^\lambda_{\sigma\nu}\Gamma^\sigma_{\mu\rho}\ .$$

From the Riemannian curvature we construct the Ricci tensor

$$R_{\mu\nu}\ =\ R^\lambda{}_{\mu\lambda\nu}\ ,$$

and the Ricci scalar

$$R\ =\ g^{\mu\nu}R_{\mu\nu}\ .$$

In a space-time devoid of matter Einstein's equations take the form

$$R_{\mu\nu}\ =\ 0\ ,$$

being ten second order partial differential equations for the metric coefficients $g_{\mu\nu}(x)$. If matter is present this contributes to gravity through its stress-energy tensor, $T_{\mu\nu}$, and Einstein's equations are

$$R_{\mu\nu} - \tfrac{1}{2}g_{\mu\nu}R\ =\ -kT_{\mu\nu}$$

where $k = 8\pi G/c^4$.

Bibliography

1. *Aristotelian Dynamics*

The quotations of Aristotle come from:
    Aristotle, *The Physics*, translated by P.H. Wickteed and F.M. Cornford, The Loeb
        Classical Library, W. Heinemann Ltd., vol. I: 1957, vol. II: 1952.
For remarks concerning the reconstruction of Aristotelian space-time structure in modern
mathematical terms, see:
    Ehlers, J., 1973, The Nature and Structure of Spacetime, in: *The Physicist's Con-*
        ception of Nature, ed. by J. Mehra, D. Reidel Pub. Co., 1973, 71-91.
    Penrose, R., 1968, Structure of Space-Time, in: *Battelle Recontres* 1967, ed. by
        C.M. DeWitt and J.A. Wheeler, W.A. Benjamin Inc., 121-235.
For the mathematical apparatus used throughout this book, our standard references are:
    Wallace, A.H., 1968, *Differential Topology – First Steps*, W.A. Benjamin Inc.
    O'Neill, 1966, *Elementary Differential Geometry*, Academic Press.
    Parts I and II of Misner, C.W., Thorne, K.S., Wheeler, J.A., 1973, *Gravitation*,
        W.H. Freeman and Co.
At a more advanced level:
    Hocking, J.G., Young, G.S., 1961, *Topology*, Addison-Wesley Pub. Co.
    Bishop, R.L., Crittendon, R.J., 1964, *Geometry of Manifolds*, Academic Press.
    Bishop, R.L., Goldberg, S.J., 1968, *Tensor Analysis on Manifolds*, Macmillan.
    Hicks, N.J., 1965, *Notes on Differential Geometry*, D. van Nostrand Co. Inc.
    Kobayashi, S., Nomizu, K., 1963, *Foundations of Differential Geometry*, John Wiley
        and Sons.
And in a very concise (but clear) manner the second chapter of the monograph:
    Hawking, S.W., Ellis, G.F.R., 1973, *The Large Scale Structure of Space-Time*,
        Cambridge, University Press.
For an introduction to Zeno's paradox see:
    Grünbaum, A., 1967, *Modern Science and Zeno's Paradoxes*, Wesleyan University Press.
For the historical development of the concept of relativity from antiquity to the present
day see:
    Tonnelat, M.A., 1971, *Histoire du principe de relativity*, Ed. Flammarion.
Other references are:
    Dedekind, R., 1901, *Essays on the Theory of Numbers*, translated by W.W. Beman,
        Open Court, London.
    Dirac, P.A.M., 1973, New Ideas of Space and Time, Naturwissenschaften, *60*, 528-531.
    Milne, E.A., 1935, *Relativity, Gravitation and the World Structure*, Clarendon
        Press, Oxford.

2. *Copernican Kinematics*

Quotations of Copernicus are taken from:
    Nicolaus Copernicus, *On the Revolutions of the Heavenly Spheres*, translated by
        J.F. Dobson and S. Brodetsky, Published originally as 'Occasional Notes of
        the Royal Astronomical Society', No. 10, 1947. Preface and Book I reprinted
        in: Theories of the Universe, ed. by M.K. Munitz, The Free Press, New York;
        Collier-Macmillan Ltd., London, 1965, 149-173.

3. *Newtonian Dynamics*

Quotations of Newton are from:
    Sir Isaac Newton's *Mathematical Principles of Natural Philosophy*, translated by
        Andrew Motte, University of California Press, 1962.
For the reconstruction of Newtonian space-time structure in modern terms see the works of
Ehlers and Penrose quoted in the references to Chapter 1, and:
    Trautman, A., 1964, Foundations and Current Problems of General Relativity, in:
        Lectures on General Relativity (Brandeis Summer Institute in Theoretical
        Physics), Prentice-Hall, Inc. (Chapter 5).
    Trautman, A., 1966, Comparison of Newtonian and Relativistic Theories of Space-
        Time, in: *Perspectives in Geometry*, ed. by B. Hoffman, Indiana Univ. Press.

Other references:
Thom, R., 1975, *Structural Stability and Morphogenesis: An Outline of a General Theory of Models*, Benjamin.
Dirac, P.A.M., 1973, Development of the physicists' conception of nature, in: *The Physicists' Conception of Nature*, ed. by J. Mehra, D. Reidel.

4.   *Critiques of Newtonian Dynamics*

Quotations of Leibniz and Clarke are taken from:
G.W. Leibniz, *Philosophical Papers and Letters*, ed. by L.E. Loemker, D. Reidel Publishing Co., 1969.
Certain physical aspects of the Leibniz-Clarke polemics, are dealt with in:
Erlichson, H., 1967, The Leibniz-Clarke Controversy: Absolute versus Relative Space-Time, American Journal of Physics, *35*, 89-98.
Heller, M., Staruszkiewicz, A., 1975, A Physicist's View on the Polemics between Leibniz and Clarke, Organon, *11*, 205-213.
Quotations of Berkeley, where indicated, are from:
The Works of George Berkeley, ed. by Alexander Campbell Fraser, Oxford, Clarendon Press, 1901 (four volumes).
Quotations of Mach are from:
Ernst Mach, *The Analysis of Sensations*, translated by C.M. Williams, Dover Pub. Inc., 1959.
Ernst Mach, *The Science of Mechanics:* Account of its Development, translated by T.J. McCormack, The Open Court Pub. Co., 1960.
See also:
Bunge, M., 1966, Mach's Critique of Newtonian Mechanics, *34*, 383-596.
Heller, M., 1975, The Influence of Mach's Thought on Contemporary Relativistic Physics, Organon, *11*, 271-283.
Other references to Chapter 4 are:
Augustynek, Z., 1970, Wtasnos'ci czasu (The Properties of Time), PWN - Warsaw.
Barbour, J.B., Bertotti, B., 1977, Gravity and Inertia in a Machian Framework, Nuovo Cim., *38B*, 1-27.
Popper, K.R., 1954, A Note on Berkeley as Precursor of Mach, The British Journal for the Philosophy of Science, *4*, 26-36.
Russell, B., 1903, (first edition), *The Principles of Mathematics*, G. Allen and Unwin Ltd., 1964.
Whitrow, G.J., 1954, Berkeley's Philosophy of Motion, ibid., 37-45.

5.   *The Space-Time of Classical Dynamics*

The geometric approach to classical dynamics is given in Ehlers (1973), Penrose (1968), and Trautman (1964, 1966), quoted in Chapters 1 and 3. Sect. 5.3 of Trautman's work (1964) is particularly useful. See also Sect. 12.2 of Misner, Thorne, Wheeler (1973) referred to in Chapter 1. References:

Cartan, E., 1923, Sur les varietes a connexion affine et la theorie de la relativite generalisee (premiere partie), Ann. Ecole Norm. Sup., *40*, 325-412.
Cartan, E., 1924, Sur les varietes a connexion affine et la theorie de la relativite generalisee (suite), Ann. Ecole Norm. Sup., *41*, 1-25.
In Sect. 5.3 we have touched the problem of the 'arrow of time'; for further reading see, for instance:
Davies, P.C.W., 1974, *The Physics of Time Asymmetry*, Surrey University Press, Tutertext Publishing Ltd.
Ellis, G.F.R., Sciama, D.W., 1972. Global and Non-Global Problems in Cosmology, in: *General Relativity (Papers in Honour of J.L. Synge)*, ed. by L. O'Raifeartaigh, Clarendon Press, Oxford, 35-59.
Heller, M., 1975, Global Time Problem in Relativistic Cosmology, Ann. Soc. Sci. Bruxelles, *89*, 522-532.
Lange, L., 1885, Uber die wissenschaftliche Fassung des Galileischen Beharrungs-gesetzes, Ber. kgl. Ges. Wiss., Math.-Phys. Kl., 333.
Layzer, D., 1971, Cosmology and the Arrow of Time, in: *Vistas in Astronomy*, ed. by A. Beer, Pergamon Press, 279-287.
Layzer, D., 1976, The Arrow of Time, Astrophys. J., *206*, 559-569.
Reichenbach, H., 1956, *The Direction of Time* (ed. by M. Reichenbach), University of California Press.

6. *Classical Space-Time in the Presence of Gravity*

See Trautman (1964, sec. 5.4-5.6) referred to in Sect. 3 and Misner, Thorne and Wheeler (1973, Chapter 12), reference in Chap. 1.

Grünbaum, A., 1957, The Philosophical Retention of Absolute Space in Einstein's Theory of Relativity, The Philosophical Review, *66*, 525-534.

Will, C.M., 1974, The theoretical tools of experimental gravitation in: *Experimental Gravitation: Proceedings of Course 56 of the International School of Physics 'Enrico Fermi'*, ed. by B. Bertotti, Academic Press, 1-110.

Will, C.M., 1979, The confrontation between gravitation theory and experiment in: *General Relativity: An Einstein Centenary Survey*, ed. by S.W. Hawking and W. Israel, Cambridge Univ. Press, 24-89.

7. *The Space-Time of Special Relativity*

The original paper of Albert Einstein on special relativity is:
Zur Elektrodynamik bewegter Körper, Annalen der Physik, *17*, 1905; English translation: On theElectrodynamic of Moving Bodies, in: *The Principle of Relativity* - A Collection of Original papers on the special and General Theory of Relativity, Dover Publications, Inc., 1952, 35-65.

The concept of space-time was introduced by Herman Minkowski in an address delivered at the 80th Assembly of German Natural Scientists and Physicists, at Cologne, 21 Sept. 1908. The English translation is the chapter Space and Time in: *The Principle of Relativity*, 73-91.

One of the numerous introductions to special relativity is:
Taylor, E.G., Wheeler, J., 1966, *Spacetime Physics*, W.H. Freeman.

Other references:
Bondi, H., 1969, *Assumption and Myth in Physical Theory*, Cambridge Univ. Press.
Marzke, R.F. and Wheeler, J.A., 1964, Gravitation and Geometry in *Gravitation and Relativity*, ed. by H.-Y. Chiu and W.F. Hoffman, Bejamin, 40-64.

8. *The Space-Time of General Relativity*

The general theory of relativity in a full form was formulated by Einstein in the paper:
Die Grundlage der allgemeinen Relativitätstheorie, Annalen der Physik, *49*, 1916; English translation: "The Foundation of the General Theory of Relativity", in: *The Principle of Relativity*, Dover Publ. Inc., 1952, 109-164.

For Einstein's further development of the theory, see, for example:
Einstein, A., 1955, *The Meaning of Relativity*, Princeton Univ. Press.

Misner, Thorne, and Wheeler, *Gravitation* (1973) (loc. cit. chapter 1) provides a modern introduction to general relativity for readers with a background in physics, as does:
Weinberg, S., 1972, Gravitation and Cosmology, *Principles and Applications of General Relativity*, John Wiley and Sons Inc.

A beautiful presentation of the mathematical structure of relativistic space-time is given by:
Schrödinger, E., 1950 (1st edition), *Space-Time Structure*, Cambridge Univ. Press.

An account of recent work at an advanced level is given in:
Hawking, S.W., Ellis, G.F.R., 1973, *The Large Scale Structure of Space-Time*, Cambridge University Press.

A selection from among a huge number of other references is:
Anderson, J.L., 1967, *Principles of Relativity Physics*, Academic Press, which presents the evolution and the meaning of the concept of relativity, in precise terms.

Rindler, W., 1977, *Essential Relativity - Special, General and Cosmological* (second edition), Springer-Verlag. This is an introductory textbook for the theory of relativity; the presentation concentrates on Mach's Principle, and the absolute-relative space problem.

Sachs, R.K., Wu, H., 1977, *General Relativity for Mathematicians*, Springer Verlag. At an advanced mathematical level. General relativity is presented as a 'trivial' interpretation of some consequences of semi-Riemannian manifold geometry.

Other references are:
Dicke, R.H., 1964, *The Theoretical Significance of Experimental Relativity*, Blackie. Most of this book consists of appendices (twelve of them) containing reprints of the original papers of Dicke and of Brans and Dicke; the book itself is a corrected version of lectures entitled ' Experimental Relativity' delivered at the Les Houches Summer School of Theoretical Physics, 1963, and published

235

    in the proceedings volume: *Relativity, Groups and Topology*, ed. by C.
    DeWitt and B. DeWitt, Gordon and Breach, 163-313.

Ehlers, J. (editor), 1979, *Isolated Gravitating Systems in General Relativity*, Pro-
    ceedings of the International School of Physics 'Enrico Fermi', Course LXVII.
    Academic Press.

Hafele, J.C., Keating, R.E., 1972, Around-the-world atomic clocks: predicted rela-
    tivistic time gains, Science, *177*, 166-168; Observed relativistic time
    gains, ibid., 168-170.

Pound, R.V., Rebka, G.A., 1960, Apparent weight of photons, Phys. Rev. Lett. *4*, 337-
    341.

Will, C.M., 1974, 1979, references in Chapter 6.

9.   *Solutions and Problems in General Relativity*

Banerji, S., 1968, Homogeneous Cosmological Models without shear, Prog. Theor. Phys.
    39, 365-371.

Bass, L., Pirani, F.A.E., 1955, On the Gravitational Effects of Distant Rotating
    Masses, Phil. Mag. *46*, 850-856.

Bianchi, L., 1918, Lezioni sulla teoria dei gruppi continui finiti transformazioni,
    Spoerri, Pisa.

Bondi, H., *Cosmology* (1960), Cambridge University Press.

Birkhoff, G.D., 1923, *Relativity and Modern Physics*, Harvard Univ. Press, Cambridge,
    Mass.

Brans, C.H., 1962, Mach's Principle and the Locally Measured Gravitational Constant
    in General Relativity, Phys. Rev., *125*, 388-396.

Brill, D.R., Cohen, J.M., 1966, Rotating Masses and Their Effect on Inertial Frames,
    Phys. Rev., *143*, 1011-1015.

Carter, B., 1971, Causal Structure in Space-Time, General Relativity and Gravitation,
    1, 349-391.

Carter, B., 1979, The general theory of the mechanical, electromagnetic and thermo-
    dynamics properties of black holes, in: *General Relativity: An Einstein
    Centenary Survey*, ed. by S.W. Hawking and W. Israel, Cambridge Univ. Press,
    294-369.

Clarke, C.J.S., Schmidt, B.G., 1977, Singularities: the State of the Art, General
    Relativity and Gravitation, *8*, 129-137.

Cohen, J.M., Brill, D.R., 1968, Further Examples of 'Machian' Effects of Rotating
    Bodies in General Relativity, Nuovo Cim., *56B*, 209-219.

De Sitter, W., 1917, On the Relativity of Inertia: Remarks concerning Einstein's
    Latest Hypothesis, Proc. Kin. Ned. Akad. Wet., *19*, 1217-1225.

Einstein, A., 1916, Die Grundlage der allgemeinen Relativitätstheorie, Ann. Phys.,
    49 (see ref. to Chapter 8).

Einstein, A., 1917, Kosmologishe Betrachtungen zur allgemeinen Relativitätstheorie,
    Sitzungster. Preuss. Akad. Wiss. *1*, 142-152, English version: Cosmological
    Considerations on the General Theory of Relativity, in: *The Principle of
    Relativity*, Dover Publ. Inc., 175-188.

Einstein, A., 1918, Prinzipielles zur allgemeinen Relativitätstheorie, Ann. Phys.,
    55, 241-244.

Einstein, A., 1955, *The Meaning of Relativity* (5th edition), Princeton Univ. Press.

Ellis, G.F.R., Schmidt, B.G., 1977, Singular Space-Times, General Relativity and
    Gravitation, *8*, 915-953.

Estabrook, F.B., Wahlquist, H.D., Behr, C.G., 1968, Dyadic Analysis of Spatially
    Homogeneous World-Models, J. Math. Phys., *9*, 497-504.

Friedman, A., 1922, Über die Krümmung des Raumes, Zeitschr. für Phys., *10*, 377-386.

Friedman, A., 1924, Über die Möglichkeit einer Welt mit Konstanter negativer Krümm-
    ung des Raumes, Zeitschr. für Phys., *21*, 236-332.

Gödel, K., 1949, An Example of a New Type of Cosmological Solution of Einstein's
    Field Equations of Gravitation, Rev. Mod. Phys., *21*, 447-450.

Gödel, K., 1952, Rotating Universes in General Relativity Theory, Proceedings of
    the International Congress of Mathematicians, 1950, Cambridge, Mass., vol. 1,
    175-181.

Hawking, S.W., 1968, The Existence of Cosmic Time Functions, Proc. Roy. Soc., *A308*,
    433-435.

Hawking, S.W., 1971, Stable and Generic Properties in General Relativity, General
    Relativity and Gravitation, *1*, 393-400.

Heller, M., 1975, Rotating Bodies in General Relativity, Acta Cosmologica, *3*, 97-
    107.

Hönl, H., Maue, A.W., 1956, Über das Gravitationsfeld rotierenden Massen, Zeitschr.
     für Phys., *144*, 152-167.
Kerr, R.P., 1963, Gravitational Field of a Spinning Mass as an Example of Alge-
     braically Special Metrics, Phys. Rev. Lett., *11*, 237-238.
King, A.R., 1972, Thesis, University of Cambridge (unpublished).
King, A.R., Ellis, G.F.R., 1973, Tilted Homogeneous Cosmological Models, Commun.
     math. Phys., *31*, 209-242.
Kruskal, M.D., 1960, Maximal Extension of Schwarzschild Metric, Phys. Rev., *119*,
     1743-1745.
Lausberg, A., 1969, Coriolis Effects in the Einstein Universe, Astronomy and Astro-
     physics, *3*, 150-155.
Lausberg, A., 1971, On the Inertial Effects Induced by a Shell of Finite Thickness,
     Bulletin de l'Academie Royale de Belgique (Classe des Sciences), *57*, 125-153.
Lens, J., Thirring, H., 1918, Über den Einfluss der Eigenrotation der Zentralkörper
     auf die Bewegung der Planeten und Monde nach der Einsteinschen Gravitations-
     theorie, Phys. Zeitschr. *19*, 156-163.
Lemaitre, G., 1927, Un Univers Homogene de Masse Constante et de Rayon Croissant
     rendant Compte de la Vitesse Radiale des Nebuleuses Extra-Galactiques, Ann.
     Soc. Sci. Brux., *47A*, 49-59. English translation: A Homogeneous Universe
     of Constant Mass and Increasing Radius accounting for the Radial Velocity
     of Extra-galactic Nebulae, Mon. Not. R. astr. Soc., *41*, 483-490 (1931).
Lemaitre, G., 1933, L'Universe en Expansion, Ann. Soc. Sci. Brux., *62A*, 51-85.
MacCallum, M.A.H., Ellis, G.F.R., 1970, A class of Homogeneous Cosmological Models.
     II. Observations. Comm. math. Phys., *19*, 31-64.
Misner, W., Thorne, K.S., Wheeler, J.A., 1973, *Gravitation*, W.H. Freeman.
North, J.D., 1965, *The Measure of the Universe*, Oxford, Clarendon Press.
Ozsvath, I., Schücking, E., 1962, Finite Rotating Universe, Nature, *193*, 1168-1169.
Penrose, R., 1974, Singularities in Cosmology, in: Confrontation of Cosmological
     Theories with Observational Data (IAU Symposium No. 63), ed. by M.S. Longair,
     Reidel.
Robertson, H.P., 1935, Kinematics and World Structure, Astrophys. J., *82*, 248-301.
Robertson, H.P., 1936, Kinematics and World Structure, Astrophys. J., *83*, 187-201;
     257-271.
Schwarzschild, K., 1916, Über das Gravitationsfeld eines Massenpunktest nach der
     Einsteinschen Theorie, Sitzber. Deut. Akad. Wiss. Berlin, KL. Maths-Phys.-
     Tech., 189-196.
Simpson, M., Penrose, R., 1973, Internal instability in a Reisener-Nordstrom black
     hole, Int. J. Theor. Phys., *7*, 183.
Teyssandier, P., 1972a, Sur le champ gravitationnel des couches spheriques simples
     tournantes a l'approximation lineaire, Lett. Nuovo Cim., *5*, 359-365.
Teyssandier, P., 1972b, On the Precession of Locally Inertial Systems in the Neigh-
     bourhood of a Rotating Sphere, Lett. Nuovo Cim., *5* (ser. 2), 1038-1043.
Thirring, H., 1918, Über die Wirkung rotierenden ferner Massen in der Einsteinschen
     Gravitationstheorie, Phys. Zeitschr., *19*, 33-39.
Thirring, H., 1921, Über die Wirkung rotierenden ferner Massen in der Einsteinschen
     Gravitationstheorie, Phys. Zeitschr., *22*, 29-30.
Walker, A.G., 1936, On Milne's theory of World-Structure, Proc. London Math. Soc.,
     42, 90-127.
Weyl, H., 1924, Massenträgheit und Kosmos: Ein Dialog, Naturwissenschaften, *12*,
     197-204.
Zel'dovitch, Ya. B., 1967, Cosmological Constant and Elementary Particles, Zh.
     Eksp. Teor. Fiz., Pis'ma, *6*, 883-884. English Version in: Sov. Phys. -
     JETP Lett., *6*, 313-317.
Zel'dovitch, Ya. B., Novikov, I.D., 1967, *Relativistic Astrophysics* (Vol. 1,
     Chicago University Press.
In Sect. 9.5 we touched the so-called cosmological problem. Modern cosmology is a rapidly
developing subject which is increasingly becoming an integral part of our knowledge of
physical reality. Introductory texts are:
Bondi, H., 1960, *Cosmology*, Cambridge University Press. Fundamental concepts of
     cosmology and main cosmological theories are clearly presented here.
Sciama, D.K., 1971, *Modern Cosmology*, Cambridge University Press, complementary
     to Bondi in stressing observational aspects and newer results are included.
Rowan-Robinson, M., 1977, *Cosmology*, Oxford, Clarendon Press - stresses the obser-
     vational aspects of modern cosmology.
Landsberg, P.T., Evans, D.A., 1977, *Mathematical Cosmology*, Oxford, Clarendon
     Press - the theoretical framework of relativistic cosmology is introduced
     via relatively simple calculations, based mainly on the Newtonian equations.

Physical cosmology is emphasised also in:

Raine, D.J., 1981, *The Isotropic Universe*, Adam Hilger.

Weinberg, S., 1977, *The First Three Minutes*, Andre Deutsch Ltd. This is a popular book of particular value. Physical processes, which according to the "standard" picture of the cosmic evolution took place in the very early universe, are presented.

At a more advanced level are:

Peebles, P.J.E., 1971, *Physical Cosmology*, Princeton Univ. Press - observational basis of modern cosmology, physical processes within the "standard" cosmological model.

Ryan, M.P., Shepley, L.C., 1975, *Homogeneous Relativistic Cosmologies*, Princeton Univ. Press - homogeneous but not necessarily isotropic world models, symmetries of cosmological models (Bianchi classification), singularities, Hamiltonian cosmology.

A concise account is given in:

Ellis, G.F.R., Relativistic Cosmology, in: *General Relativity and Cosmology* (Proc. of 47 Intern. School of Phys. Enrico Fermi), ed. by R.K. Sachs, Academic Press, 1971, 104-182. Or in a slightly modified version: Relativistic Cosmology, in: Cargese Lectures in Physics, vol. 6, ed. by E. Schatzman, Gordon and Breach, 1-60.

For a history of modern cosmology see:

North, J.D., *The Measure of the Universe*, Oxford, Clarendon Press, 1965.

Merleau-Ponty, J., 1965, *Cosmologie du XXe siecle*, Ed. Gallimard.

The full list of the original papers on cosmology for the period 1917-1932 may be found in the article:

Robertson, H.P., Relativistic Cosmology, Rev. Mod. Phys., *5*, 1933, 62-90.

A complete list for the period 1933-1940 appears in:

Heckmann, O., 1968, Theorien der Kosmologie, Springer Verlag.

The status of Mach's Principle in contemporary relativistic physics is reviewed in the following papers:

Goenner, H., 1970, Mach's Principle and Einstein's Theory of Gravitation, Boston Studies in the Philosophy of Science, Vol. VI, D. Reidel Pub. Co., 200-215.

Reinhardt, M., 1973, Mach's Principle - A Critical Review, Zeitschr. für Naturforschung, *28a*, 529-537.

10.   *Mach's Principle and the Dynamics of Space-Time*

Arnowitt, R.L., Deser, S., Misner, C.W., 1962, The Dynamics of General Relativity, in: *Gravitation: An Introduction to Current Research*, ed. by L. Witten, John Wiley and Sons, 227-265.

Bruhat, Y., 1926, The Cauchy Problem, in: *Gravitation: An Introduction to Current Research*, 130-168.

DeWitt, B.S., 1970, Spacetime as a Sheaf of Geodesics in Superspace, in: *Relativity* (Proc. Relativity Conf. in the Midwest, 1969), ed. by M. Carmeli, S.I. Fickler and L. Witten, Plenum Press, 359-374.

Dirac, P.A.M., 1959, Fixation of Coordinates in the Hamiltonian Theory of Gravitation, Phys. Rev., *114*, 924-930.

Dirac, P.A.M., 1964, *Lectures on Quantum Mechanics*, Yeshiva Univ., New York.

Fischer, A.E., 1970, The Theory of Superspace, in: *Relativity* (Proc. Relativity Conf. in the Midwest, 1969), 303-357.

Fischer, A.E., Marsden, J.E., 1979, The Initial Value Problem and the Dynamical Formulation of General Relativity, in: *General Relativity, An Einstein Centenary Survey*, ed. by S.W. Hawking and W. Israel, Cambridge Univ. Press, 138-211.

Fletcher, J.G., 1962, Geometrodynamics, in: *Gravitation: An Introduction to Current Research*, ed. by L. Witten, Wiley, 412-437.

Gott, J.R. III, Gunn, J.E., Schramm, N., Tinsley, M., 1974, Astrophys. J., *194*, 543-553.

Hartle, J.B., 1960, A.B. Thesis, Princeton.

Hilbert, D., 1951, Die Grundlagen der Physik, Konigl. Gesell. Wiss. Göttingen, Nachr., Math-Phys. Kl, 395-407.

Isaacson, R.A., 1968, Gravitational Radiation in the Limit of High Frequency, I: The Linear Approximation and Geometrical Optics, Phys. Rev., *166*, 1263-1271; II: Nonlinear Terms and the Effective Stress Tensor, Phys. Rev., *166*, 1272-1280.

Isenberg, J., 1974, Initial Value Formulation of Mach's Principle, Bull. Am. Phys. Soc., *19*, 508.

238

Lichnerowicz, A., 1955, Theories relativistes de la gravitation et de l'electro-
     magnetisme, Masson et Cie.
O'Murchadha, N., York, J.W., 1974, Initial Value Problem of General Relativity, I:
     General Formulation and Physical Interpretation, Phys. Rev., *10D*, 428-436;
     II: Stability of Solutions of the Initial-Value Equations, Phys. Rev., *10D*,
     437-446.
Stern, M.D., 1967, Regularity of the Topology of Superspace, Princeton Senior
     Thesis A.B.
Wheeler, J.A., 1955, Geons, Phys. Rev., *97*, 511-536.
Wheeler, J.A., 1962a, *Geometrodynamics* (Topics on Modern Physics, vol. I), Academic
     Press.
Wheeler, J.A., 1964a, Geometrodynamics and the Issue of the Final State, in: *Rela-
     tivity, Groups and Topology* (Les Houches Summer School of Theor. Phys. 1963),
     ed. by C. DeWitt and B. DeWitt, Gordon and Breach, 315-520.
Wheeler, J.A., 1964b, Gravitation as Geometry - II, in: *Gravitation and Relativity*,
     ed. by H.Y. Chiu and W.F. Hoffman, W.A. Benjamin Inc., 65-89.
Wheeler, J.A., 1964c, Mach's Principle as Boundary Condition for Einstein's Equa-
     tions, in: *Gravitation and Relativity*, 303-349.
Wheeler, J.A., 1967, Three-Dimensional Geometry as a Carrier of Information about
     Time, in: *The Nature of Time*, ed. by T. Gold, Cornell Univ. Press, 90-110.
Wheeler, J.A., 1968, Superspace and the Nature of Quantum Geometrodynamics, in:
     Battelle Rencontres, ed. by C.M. DeWitt and J.A. Wheeler, W.A. Benjamin Inc.,
     242-307.
York, J.W., 1971, Gravitational Degrees of Freedom and the Initial-Value Problem,
     Phys. Rev. Lett., *26*, 1656-1658.
York, J.W., 1972, Role of Conformal Three Geometry in the Dynamics of Gravitation,
     Phys. Rev. Lett., *28*, 1082-1085.
York, J.W., 1973, Conformally Invariant Orthogonal Decomposition of Symmetric Ten-
     sors on Riemannian Manifolds and the Initial-Value Problem of General Rela-
     tivity, J. Math. Phys., *14*, 456-464.
For a concise presentation of Wheeler's 'plan of General Relativity' see Chapter 21 of
Gravitation by Misner, Thorne and Wheeler.

11. *Theories of Inertial Mass*

Beltran-Lopez, V., Robinson, H.G., Hughes, V.W., 1961, Improved Upper Limit for the
     Anisotropy of Inertial Mass from a Nuclear Magnetic Resonance Experiment,
     Bull. Am. Phys. Soc., *6*, 424.
Brans, C., Dicke, R.H., 1961, Mach's Principle and Relativistic Theory of Gravita-
     tion, Phys. Rev., *124*, 925-935.
Brans, C., 1962, Mach's Principle and a Relativistic Theory of Gravitation II,
     Phys. Rev., *125*, 2194-2201.
Carrelli, A., 1959, On Mach's Principle, Nuovo Cim., *13*, 853-856.
Cocconi, G., Salpeter, E.E., 1958, A Search for Anisotropy of Inertia, Nuovo Cim.,
     10, 646-651.
Cocconi, G., Salpeter, E.E., 1960, Upper Limit for the Anisotropy of Inertia from
     the Mössbauer Effect, Phys. Rev. Lett., *4*, 176-177.
Deser, S., Pirani, F.A.E., 1965, Critique of a New Theory of Graviation, Proc. Roy.
     Soc., *A288*, 133-145.
Dicke, R.H., 1957a, Principle of Equivalence and the Weak Interactions, Rev. Mod.
     Phys., *29*, 355-362.
Dicke, R.H., 1957b, Gravitation without a Principle of Equivalence, Rev. Mod. Phys.
     29, 363-376.
Dicke, R.H., 1961, Experimental Tests of Mach's Principle, Phys. Rev. Lett., *7*,
     359-360.
Dicke, R.H., 1962a, Mach's Principle and Invariance under Transformation of Units,
     Phys. Rev., *125*, 2163-2167.
Dicke, R.H., 1962b, Mach's Principle and Equivalence, in: *Evidence for Gravita-
     tional Theories* (Intern. School of Phys. Enrico Fermi Course 20), ed. by
     C. Møller, 1-49.
Dicke, R.H., 1964, *The Theoretical Significance of Experimental Relativity*,
     Blackie - see reference in Chapter 8.
Dirac, P.A.M., 1938, New Basis for Cosmology, Proc. Roy. Soc., *A165*, 199-208.
Dirac, P.A.M., 1973a, Long Range Forces and Broken Symmetries, Proc. Roy. Soc.,
     A333, 403-418.
Dirac, P.A.M., 1973b, New Ideas of Space and Time, Naturwissenschaften, *60*, 528-531.

Drever, R.W.P., 1961, A Search for Anisotropy of Inertial Mass using a Free Precession Technique, Phil. Mag., *6*, 683-687.

Hoyle, F., 1975, On the Origin of the Microwave Background, Astrophys. J., *198*, 683-687.

Hoyle, F., Narlikar, J.V., 1964, Time Symmetric Electrodynamics and the Arrow of Time in Cosmology, Proc. Roy. Soc., *277A*, 1-23.

Hoyle, F., Narlikar, J.V., 1965, A New Theory of Gravitation, Proc. Roy. Soc., *282A*, 191-207.

Hoyle, F., Narlikar, J.V., 1966, A Conformal Theory of Gravitation, Proc. Roy. Soc., *294A*, 138-148.

Hoyle, F., Narlikar, J.V., 1974, *Action at a Distance in Physics and Cosmology*, Freeman.

Hughes, V.W., 1964, Mach's Principle and Experiments on Mass Anisotropy, in: *Gravitation and Relativity*, ed. by Hong-Yee Chiu and W.F. Hoffman, W.A. Benjamin, 106-120.

Hughes, V.W., Robinson, H.G., Beltran-Lopez, V., 1960, Upper Limit for the Anisotropy of Inertial Mass from Nuclear Resonance Experiments, Phys. Rev. Lett., *4*, 342-344.

Infeld, L., Schild, A., 1945, A New Approach to Kinematic Cosmology, Phys. Rev., *68*, 250-272.

Jordan, P., 1955, *Schwerkraft und Weltall*, F. Viewig.

Jordan, P., 1959, Zum gegenwärtigen Stand der Diracschen kosmologischen Hypothesen, Zeitschr. für Phys., *157*, 112-121.

Jordan, P., 1973, The Expanding Earth, in: *The Physicist's Conception of Nature*, ed. by Jagdish Mehra, D. Reidel, 60-70.

Nordtvedt, K., 1977, Theoretical Significance of Present-Day Gravity Experiments, *Proc. of the First Marcel Grossman Meeting on General Relativity, Trieste, 1975*, North Holland, 539-544.

Partridge, R.B., 1977, Observational Cosmology, *Proc. of the First Marcell Grossman Meeting on General Relativity, Trieste, 1975*, North Holland, 617-648.

Reinhardt, M., Eichendorf, W., 1977, How Constant are Fundamental Physical Quantities?, Zeitschr. für Naturforsch., *32a*, 532-537.

Scherwin, C.W., Frauenfelder, H., Garwin, E.L., Lüscher, E., Margulies, S., Peacock, R.N., 1960, Search for the Anisotropy of Inertia using the Mössbauer Effect in Fe^{57}, Phys. Rev., *4*, 399-401.

Taylor, J.H., Fowler, L.A., McCulloch, P.M., 1979, Measurements of General Relativistic Effects in the Binary Pulsar PSR1913+16, Nature, *277*, 437-440.

Toton, E., 1968, Report on the Investigation of Mach's Principle in the Scalar-Tensor Theory of Gravity, Phys. Rev. Lett., *21*, 1401-1403.

Toton, E., 1970, Scalar-Tensor Theory in Canonical Form I: the Question of Mach's Principle, J. Math. Phys., *11*, 1713-1724.

Tetrode, H., 1922, Über den Wirkungszusammentang der Welt. Eine Erweiterung der klassischen Dynamik, Zeitschr. für Phys., *10*, 317-328.

Wheeler, J.A., Feynman, R.P., 1945, Interaction with the Absorber as the Mechanism of Radiation, Rev. Mod. Phys., *17*, 157-181.

Wheeler, J.A., Feynman, R.P., 1949, Classical Electrodynamics in Terms of Direct Inter-Particle Action, Rev. Mod. Phys., *21*, 425-434.

12.   *Integral Formulations of General Relativity*

Altshuler, B.L., 1961, Integral Form of the Einstein Equations and a Covariant Formulation of the Mach Principle (in Russian), Zh. Exper. Theor. Fiz., *51*, 1143-1150; English translation: Soviet Phys. JETP, *24*, 766 (1967); *28*, 687 (1969).

Barrow, J.D., 1977, The Homogeneity and Isotropy of the Universe, Mon. Not. R. astr. Soc., *181*, 719-727.

Bondi, H., 1947, Spherically Symmetric Models in General Relativity, Mon. Not. R. astr. Soc., *107*, 410.

Carter, B., 1974, Large Number Coincidences and Anthropic Principle in Cosmology, in: *Confrontation of Cosmological Theories with Observational Data* (IAU Symp. 63), ed. by M.S. Longair, D. Reidel, 291-298.

Collins, C.B., Hawking, S.W., 1973, Why is the Universe Isotropic?, Astrophys. J., *180*, 317-334.

Davidson, W., 1957, General Relativity and Mach's Principle, Mon. Not. R. astr. Soc., *117*, 212-224.

Deser, S., 1970, Self-Interaction and Gauge Invariance, General Relativity and Gravitation, *1*, 9-18.

Dicke, R.H., Peebles, P.J.E., Roll, P.G., Wilkinson, D.T., 1965, Cosmic Black-Body Radiation, Astrophys. J., *142*, 414-419.

Ehlers, J., Kundt, W., 1962, Exact Solutions of the Gravitational Field Equations, in: *Gravitation: an Introduction to Current Research*, ed. by L. Witten, John Wiley, 49-101.

Ellis, G.F.R., Sciama, D.W., 1972, Global and Non-Global Problems in Cosmology, in: *General Relativity* (Papers in Honour of J.L. Synge), ed. by L. O'Raifeartaigh Clarendon Press, Oxford, 35-59.

Gilman, R.C., 1969, A Completely Covariant Integral Theory of Inertia and Gravitation and its Machian Implications, Thesis, Princeton Univ.

Gilman, R.C., 1970, Machian Theory of Inertia and Gravitation, Phys. Rev., *2D*, 1400-1410.

Groth, E.J., Peebles, P.J.E., Seldner, M., Soneira, R.M., 1977, The Clustering of Galaxies, Scientific American, *237*, 76-98.

Kristian, J., Sachs, R.K., 1966, Observations in Cosmology, Astrophys. J., *143*, 379-399.

Lynden-Bell, D., 1967, On the Origins of Space-Time and Inertia, Mon. Not. R. astr. Soc., *135*, 413-428.

Matzner, R.A., Misner, C.W., 1972, Dissipative Effects in the Expansion of the Universe, Astrophys. J., *171*, 415-432.

Misner, C.W., 1968, The Isotropy of the Universe, Astrophys. J., *151*, 431-457.

Misner, C.W., 1969, Mixmaster Universe, Phys. Rev. Lett., *22*, 1071-1074.

Muller, R., 1978, The Cosmic Background Radiation and the New Aether Drift, Scientific American, *238*, 64-74.

Penrose, R., 1964, Conformal Treatment of Infinity, in: *Relativity, Groups and Topology* (Les Houches Summer School of Theor. Phys., 1963), ed. by C. DeWitt and B. DeWitt, Gordon and Breach, 563-584.

Penrose, R., 1968, Structure of Space-Time, in: *Battelle Recontres*, 1967, ed. by C.M. DeWitt and J.A. Wheeler, W.A. Benjamin, 121-235.

Penzias, A.A., Wilson, R.W., 1965, A Measurement of Excess Antenna Temperature at 4080 Mc/s., Astrophys. J., *142*, 419-421.

Raine, D.J., 1971, Mach's Principle in General Relativity, Thesis, Cambridge Univ.

Raine, D.J., 1975, Mach's Principle in General Relativity, Mon. Not. R. astr. Soc., *171*, 507-528.

Raine, D.J., 1981, *The Isotropic Universe*, Adam Hilger.

Sachs, R.K., Wolfe, A.M., 1967, Perturbations of a Cosmological Model and Angular Variations in the Microwave Background, Astrophys. J., *147*, 73-90.

Sciama, D.W., 1953, On the Origin of Inertia, Mon. Not. R. astr. Soc., *113*, 34-42.

Sciama, D.W., 1964, The Physical Structure of General Relativity, Rev. Mod. Phys., *36*, 463-494.

Sciama, D.W., 1971, Astrophysical Cosmology, in: *General Relativity and Cosmology*, Proc. Int. School of Theor. Phys. Enrico Fermi, Course 47, Academic Press, 183-236.

Sciama, D.W., 1974, Gravitational Waves and Mach's Principle, in: *Ondes et Radiations Gravitationelles* (Coll. Intern. C.N.R.S.), 267-282.

Sciama, D.W., Waylen, P.C., Gilman, R.C., 1969, Generally Covariant Integral Formulation of Einstein's Field Equations, Phys. Rev., *187*, 1762-1766.

Smoot, G.F., Gorenstein, M.V., Muller, R.A., 1977, Detection of Anisotropy in the Cosmic Blackbody Radiation, Phys. Rev. Lett., *39*, 898-901.

Zel'dovitch, Ya.B., 1972, Creation of Particles and Antiparticles in an Electric and Gravitational Field, in: *Magic without Magic*, ed. by J. Klander, Freeman, 277-288.

Zel'dovitch, Ya.B., Novikov, I.D., 1975, *Structure and Evolution of the Universe* (in Russian), ed. "Nauka", Moscow.

13. *Frontiers of Relativity*

Boulware, D.G., Deser, S., 1975, Classical General Relativity Derived from Quantum Gravity, Ann. Physics, *89*, 193.

Barbour, J.B., 1974, Relative-Distance Machian Theories, Nature (Phys. Sci.), *249*, 328-329 (Misprints corrected in: Nature, *250*, 606, 1974).

Barbour, J.B., Bertotti, B., 1977, Gravity and Inertia in a Machian Framework, Nuovo Cim., *38B*, 1-27.

Cartan, E., 1922, Sur une generalisation de la notion de courbure de Riemann et les espaces a torsion, Acad. Sci. Paris, Comptes Rend., *174*, 593-595.

Cartan, E., 1923, Sur le varietes a connexion affine et la theorie de la relativite generalisee, Ann. Ecole Norm. Sup., *40*, 325-412.

DeWitt, B.S., 1964, Dynamical Theory of Groups and Fields, in: *Relativity, Groups and Topology* (Les Houches Summer School of Theor. Phys. 1963), ed. by C. DeWitt and B. DeWitt, Blackie, 585-820.
DeWitt, B.S., 1967, Quantum Theory of Gravity:
    I.   The Canonical Theory, Phys. Rev., *160*, 1113-1148.
    II. The Manifestly Covariant Theory, ibid., *162*, 1195-1239.
    III. Applications of the Covariant Theory, ibid., *162*, 1239-1256.
Ehlers, J., 1973, The Nature and Structure of Spacetime, in: *Physicist's Conception of Nature*, ed. by J. Mehra, D. Reidel, 71-91.
Einstein, A., 1955, *The Meaning of Relativity* (5th edition), Princeton Univ. Press.
Fadeev, L.D., Popov, V.N., 1967, Feynman Diagrams for the Yang-Mills Field, Phys. Lett., *25B*, 29-30.
Feynman, R.P., Hibbs, A.R., 1965, *Quantum Mechanics and Path Integrals*, McGraw-Hill.
Freedman, D.Z., van Nieuwenhuizen, P., Supergravity and the Unification of the Laws of Physics, Scientific American, *238*, 126-143.
Goldoni, R., 1976, A Background Dependent Approach to the Theory of Gravitation:
    I. Quasistatic Limit, General Relativity and Gravitation, *7*, 731-741.
    II. Relativistic Gravitational Equations, ibid., *7*, 743-755.
Graves, J.C., 1971, *The Conceptual Foundations of Contemporary Relativity Theory*, the MIT Press, Cambridge, Mass.
Hawking, S.W., 1975, Particle Creation by Black Holes, in: *Quantum Gravity*, ed. by C.J. Isham, R. Penrose and D.W. Sciama, Clarendon Press, Oxford, 219-267.
Hawking, S.W., 1979, The Path Integral Approach to Quantum Gravity, in: *General Relativity, An Einstein Centenary Survey*, ed. by S.W. Hawking and W. Israel, Cambridge Univ. Press, 746-789.
Hehl, F.W., Spin and Torsion in General Relativity,
    1973, I. Foundations, General Relativity and Gravitation, *4*, 333-349.
    1974, II. Geometry and Field Equations, General Relativity and Gravitation, *5*, 491-516.
Heller, M., 1970, Mach's Principle and Differentiable Manifold, Acta Phys. Pol., *B1*, 131-138.
Jones, R.M., 1974, Mach's Principle and Quantum Theory I: A Quantum Statement of Mach's Principle, Univ. of Colorado (unpublished).
Kibble, T.W., 1961, Lorentz Invariance and the Gravitational Field, J. Math. Phys., *2*, 212-221.
Kuchowicz, B., Cosmology with Spin and Torsion:
    1975, I. Physical and Mathematical Foundations, Acta Cosmologica, *3*, 109-129.
    1976, II. Spatially Homogeneous Aligned Spin Models with the Weyssenhoff Fluid, ibid., *4*, 67-100.
Misner, C.W., 1960, Wormhole Initial Conditions, Phys. Rev., *118*, 1110-1111.
Misner, C.W., Wheeler, J.A., 1957, Classical Physics as Geometry, Ann. Phys., *2*, 525-603.
Penrose, R., 1968, Twistor Quantization and the Curvature of Space-time, Int. J. Theor. Phys., *1*, 61-99.
Penrose, R., 1975, Twistor Theory, its aims and achievements, in: *Quantum Gravity* - ed. by C.J. Isham, R. Penrose and D.W. Sciama, Clarendon Press, Oxford, 268-407.
Penrose, R., MacCallum, M.A.H., 1973, Twistor Theory: An Approach to the Quantization of Fields and Spacetime, Physics Reports, *6C*, 242-315.
Reinich, G.W., 1924, Electrodynamics in the General Relativity Theory, Proc. Nat. Acad. Sci. (USA), *10*, 124-127; 294-298.
Sciama, D.W., 1962, On the Analogy between Charge and Spin in General Relativity, in: *Recent Developments in General Relativity*, Pergamon Press - PWN, Polish Scientific Publishers, 415-439.
Schrödinger, E., 1963, *Space-Time Structure*, Cambridge Univ. Press.
Tonnelat, M.A., 1955, La theorie du champ unifie d'Einstein et quelques-uns de ses developements, Gauthier-Villars, Paris.
Tonnelat, M.A., 1966, The Principles of Electromagnetic Theory and of Relativity, D. Reidel.
Wheeler, J.A., 1964, Geometrodynamics and the Issue of the Final State, in: *Relativity, Groups and Topology* (Les Houches Summer School of Theor. Phys. 1963), ed. by C. deWitt and B. DeWitt, Blackie, 315-520.
Wheeler, J.A., 1968, Superspace and the Nature of Quantum Geometrodynamics, in: *Battelle Rencontres*, ed. by C. DeWitt and J.A. Wheeler, W.A. Benjamin, 242-307.
Wheeler, J.A., 1973, From Relativity to Mutability, in: *The Physicist's Conception of Nature*, ed. by J. Mehra, D. Reidel, 202-247.

Index

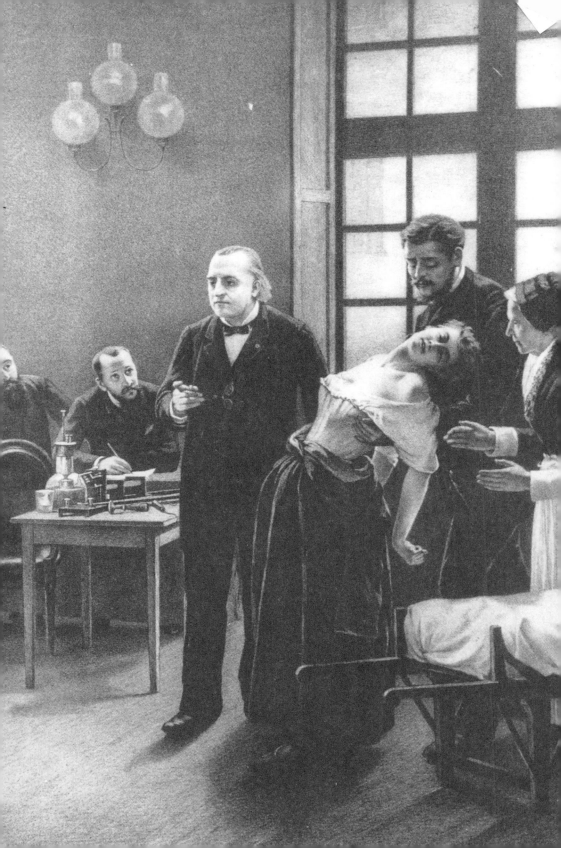